Index

Foreword

How to use this manual

Lab 1	Micropipettes
Lab 2	Dilutions

Project 1 overview: Analysis of the TAS2R32 locus—are you a taster?

Lab 3	DNA extraction principle: preparing the Lysis Buffer from human cells
Lab 4 F2F	DNA extraction from human hair p1: : collection and lysis
Lab 5 F2F	DNA extraction p2: Extraction proper and Precipitation
Lab 6 F2F	DNA extraction p3: Resuspension and Quantification
Lab 7	PCR theory
Lab 8	PCR set up and the importance of controls
Lab 9	Bioinformatics, TAS2R38 theory, and primer design
Lab 10 F2F	Polymerase Chain Reaction
Lab 11	Gel electrophoresis theory
Lab 12 F2F	Gel electrophoresis: gel preparation
Lab 13 F2F	Gel electrophoresis: run
Lab 14	Data analysis (gel) and principles of amplicon purification and DNA sequencing
Lab 15 F2F	Amplicon Purification and DNA Sequencing Prep (PTC pickup)
Lab 16	DNA sequencing analysis, Hardy Weinberg, and data discussion
Lab 17	How to Write a Lab Report

Project 2 overview: Expression analysis—Determining mRNA copy number.

Lab 18	RNA extraction in theory: differences and similarities with DNA extraction
Lab 19	Retro-transcription: breaking the dogma
Lab 20	Bioinformatics, primer design for expression analysis
Lab 21	Real time PCR theory
Lab 22	Serial dilutions
Lab 23-25	Real time data analysis
Appendix A	The cost of molecular science
References	

Foreword

To the Students:

I have many years of experience working in research laboratories, and it can be as challenging as it is fun.

It will be interesting to all of you to get a glimpse of that world, even if just to understand if this is something you could pursue for a future career. Many science careers require to be heavily involved in the lab. Researchers seeking all kinds of knowledge in the most disparate fields (cancer biology, pharmacology, environmental biology, diagnosis, prevention, etc...) look for answers by using many of the basic tools you will become familiar with during this semester.

Also, the ability to integrate knowledge from math, chemistry, physics, stats, and biology, together with developing the ability to think fast on your feet, will be a resource to all of those going toward the health professions (pre-med, pre-vet, chiropractic, physical therapy etc...)

All of you will feel a bit clumsy and disoriented in the beginning. Some will catch on quickly and fall in love with the lab work, some will not. This manual is designed to cater to the needs of both categories.

Those of you that just want to survive the lab and move on will find easy and detailed step by step instructions. Those becoming more and more passionate about the lab experience will find plenty of in-depth explanations and challenging questions.

No matter which category you will find yourself in, remember that being a scientist means to understand exactly what you are doing so that you can correct it and improve it.

That said, have fun!

How to use this manual

Each of the two projects we will carry out throughout the semester has an outline to remind you of the workflow of the experiments throughout the weeks necessary to complete the project. Because of COVID, I restructured the manual to minimize time on campus and maximize your safety.

Lab sections:

- Objective- the goal of that lab segment
- Vocabulary- a list of pertinent new words
- Background- the basic knowledge required to understand new techniques, put them in context, and learn their most common applications
- Tips- to help you executing tricky techniques correctly
- Experiments- each experiment has a specific practical goal. More experiments are typically needed to carry out one project. Projects ask a biological question. For example, we will want to find if a certain gene is expressed in your buccal cells. In order to do that we will need to perform several experiments; extract the RNA of your buccal cells, retro-transcribe it into cDNA, design appropriate expression analysis primers, carry out a real-time PCR, and analyze the results. Each experiment will, in turn, be divided in the following subsections:
    - Materials- a list of all required reagents and supplies to carry out a specific protocol
    - Experiment overview- a general outline of the steps comprised in the experiment with concise explanations justifying them
    - Set up- some experiments will require some thinking and preparation prior to execution
    - Protocols- what to do, step by step, to perform one experiment

This manual is designed to become more challenging as your skills grow. We start with a review of dilutions, which you likely learned in intro chemistry. Then you will extract DNA from your own cells. As you get more proficient and confident with your skills experiments will become more complicated, culminating in the real-time data analysis, which will take us several meetings to complete.

Throughout the manual, keywords are in **bold**, to help you pinpoint them easily.

This manual is interactive. The only way to perform lab science correctly is to understand what you are doing and why. You have to learn to start questioning everything you do and come up with your own answers. This is an essential skill you need to develop to be a successful scientist, regardless of the field. As you transition from school to the professional world, you have to switch your mentality from learning to be able to generate knowledge in order to become a leader in the scientific community. This manual helps you to think in a critical fashion by asking you a lot of questions. Take it like a game, and see if you are up to the challenge. All the answers will be discussed in class. Every time your answer is required you will see a checkmark like below

✓ Understood?

Make sure to compile all answers correctly and legibly.

Speaking up is HARD. It's easier for extroverts and harder for introverts, but it's a skill you'll need to acquire. To encourage you some answers have points associated with them. You can work on each answer solo or in teams, and everyone giving a correct answer gets points toward the lab sovereign, to be crowned at the end of the semester!

HINTS help you on the most challenging questions and exercises.

The major concepts in each chapter are underlined, for easy visualization.

You can find a list of all references at the end of the book.

Fill in the "Results" section at the end of the book. This will be important for your grade, and it will help you to write your reports.

Finally, an appendix on the 'cost of molecular science' gives you an idea of the cost of some of the reagents we are going to use during the semester.

All Material Safety Data Sheet (MSDS) of reagents are available online but will be discussed in class.

I hope you will enjoy this semester as much as I always do!

Gaia Bistulfi Amman, PhD

Micropipettes

Objective

The objective of this lab is for you to become familiar with how laboratory micropipettes work. You will get to try them, hand-on soon!

Vocabulary

- **Dispense** - to deliver the sample
- **Adjustment** - to change micropipette settings to dispense a specific volume
- **Aspirate** - to draw up the sample
- **Aliquot** - a small portion of a sample
- Volumes in a research lab can be tiny! Review the metric system prefixes **milli** (10^{-3}), **micro** (10^{-6}), **nano** (10^{-9}), **pico** (10^{-12})
- **Calibration** - ensuring that the volume on the dial of a micropipette is the volume actually drawn by the pipet
- **Stock** - the original solution from which an aliquot is drawn. If concentration calculations are involved, the stock is always the more concentrated solution

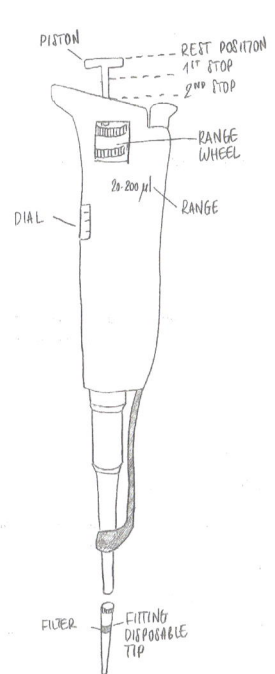

Air displacement micropipette
The range of the pipette is usually indicated on the side. The volume is set with the range wheel, while checking on the dial. Disposable filter tips, with or without filter, are used to prevent sample contamination. The piston is operated to draw and release samples.

Background

Micropipettes are precise devices used to dispense very small volumes, typically in the range of 1-1000 μl (0.001-1 mL)

We will use the most common type of micropipette; **Air Displacement Pipettes** (see figure beside). They are meant for general use with aqueous solutions. In air displacement pipettes, pushing the piston down expels a precise air volume from the disposable tip, allowing to draw the corresponding volume, from a solution. Pushing the piston down all the way (to the second stop) will release the sample.

Micropipettes operate within different ranges. Typical ranges of operation are within an order of ten like: 0.2-2 μl, 2-20 μl, 20-200 μl, 100-1000 μl. Each pipette is named after the maximum volume it can draw. For example 'p200' refers to a micropipette working within the range of 20-200 μl. Each pipette type is used with a fitting disposable tip. Learn to arrange disposable tip boxes on your bench in order of size, not to get confused. Some disposable tips come with a filter. These are used to prevent contamination from the pipette to the sample, and vice versa, in very sensitive applications like PCR, RNA based applications, and tissue culture.

TIPS

- NEVER USE A PIPET OUT OF ITS RANGE! They might get jammed and are expensive to fix. Worse, they might lose calibration
    - ✓ Why would that be worse (1 pt)?
- Always handle pipettes with great care!
- Learn to operate the piston smoothly with your thumb
- Always operate the piston SLOWLY to avoid the formation of bubbles
- Always draw solutions at an angle to avoid contamination of your samples. Never have someone else hold the tube you want to draw from!
- Learn to keep the disposable tip of your pipette just below the surface of the sample; as you aspirate the sample, smoothly move the tip down to keep it below the surface
- Once the sample is drawn you can even remove the thumb from the piston, the sample will stay in the tip (unless it has a very low viscosity, like chloroform)
- Always choose tips closer to the edge of the box to avoid contaminating other tips. If any bubbles are formed empty the tip back in the original sample and repeat
- Minimize the time a wet tip stays in the air to minimize contamination
- When pipetting several reagents into a tube, add first the bigger volume, unless otherwise indicated
- Different pipettes require different disposable tips. Set tips' boxes on your bench in order of size, not to get confused

Contamination is the enemy

Contamination means adding to your sample something that was not originally there and that might interfere with your experiment by giving you faulty results. It is of the prime importance to avoid sample contamination.

- Samples can be contaminated from you speaking over your open tubes or shedding dead skin cells into your samples
- Samples can be cross-contaminated! If your disposable tip touches one sample, dispose of it before moving to the next (unless the samples are identical)
- Filter tips are used to minimize contamination from pipettes to samples (and vice versa) in the most sensitive applications (RNA extraction and handling, PCR, tissue culture)

- <u>NEVER walk around with pipette tips filled with fluid. NEVER tilt or rest a full pipette</u>
- Always have appropriate <u>controls</u> in your experiments to check for sample contamination

Dilutions

Objective

Today's objective is for you to understand dilutions and perform pertinent calculations correctly

Vocabulary

- **To dilute** -to decrease the concentration of a solute. Concentrations are typically expressed as mass/volume and can be calculated in several ways; weight/volume, moles/volume, number of molecules/volume, percentages

- **Initial concentration and initial volume**- refer to the concentration and volume you take from your stock solution (C_1, V_1) in order to make a dilution

- **Final concentration and final volume**- refer to the concentration and volume you want (C_2, V_2) for your working solution

- **Dilution factor**- the ratio between the concentration of a stock and the concentration of the working solution (C_1/C_2).

Background

As you will soon find out, in the lab, you need quite often to prepare solutions with a specific concentration. It is very important to understand the theory of dilutions otherwise you might compromise your experiments!

Today we will learn about:

- Dilutions in general
- Strength dilutions
- Percent dilutions

Later in the semester you will learn about:

- Serial dilutions

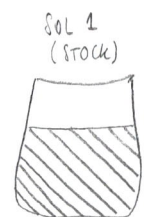

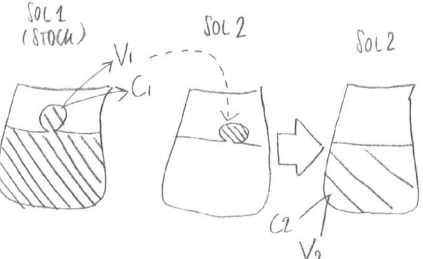

Dilutions. A stock solution has a concentration (C_1) of solute. A volume (V_1) is taken from the stock to prepare a solution at the desired concentration (C_2) in the desired volume (V_2)

When you make a dilution, you know the initial concentration of your stock (C1), the final concentration (C2) and final volume (V2) you need.

So you need to determine how much to take from your stock (V1) like shown in the figure above.

The total amount of "solute" you take out of solution 1 (your stock) depends on the volume you take out (V1) and its concentration (C1).

- ✓ For example, if solution 1 has a solute concentration of 8 $\mu g/\mu l$, and you take out 100 μl, how many μg total do you take out of solution 1? (1 pt)

If you add what you took out to pure water, the amount of solute (v1*C1) will not change, but it will be now in a bigger volume; you made a dilution. This bigger volume is V2 = V1+water=V2.

- ✓ So, if I add the 100 μL above (8 $\mu g/\mu l$) to a water amount of 200 μL, what is V2 (1 pt)? What is my new concentration, C2 (1 pt)?
- ✓ How would you generalize the calculation you just made with a formula involving C1, V1, C2, and V2 (2 pts)? C2 = ?

- ✓ You can write a formula linking C1, V1, C2, and V2 in a format that is quite easy to remember. What is it? (1 pts)

You can use this formula for most dilution calculations.

Just remember that V2 is your FINAL VOLUME (Water plus V1).

- ✓ Which one is always bigger, V1 or V2? (1 pt) C1 or C2? (1 pt)

- ✓ You have a solution of 25 mM $MgCl_2$. You need 50 μl of 10 mM $MgCl_2$. What would you do?

Strength dilutions

For some reagents the final concentration you need is always the same.

In this case the stock may be labeled as a '5x' or '10x'. This means the initial concentration C1 is 5 or 10 times, respectively, more concentrated than the final concentration you need, C2. Basically, your dilution factor $C_1/C_2 = 10$ for the 10X. The easiest way to solve these problems is to remember that, if you have a 10x, you need to dilute 10 times ($V_1 = V_2/10$), for a 25X, $V_1 = V_2/25$.

- ✓ For example, you need to make 100 μl of a solution from a 10x stock. What would you mix (1 pt)?

- ✓ Now you need to make 55 μl of 1x solution from a 2.5x stock. What would you mix (1 pt)?

Percent dilutions

Percent dilutions are used to dissolve a solid into a liquid. For example a 5% glucose solution, by definition, means you dissolve 5g of glucose into 100 mL of water (unless a different solvent is specified). To solve this type of dilutions, just set up a proportion to obtain the volume you need!

- ✓ You need to make 100 mL of an 8% NaCl solution. What do you mix (1 pt)?

- ✓ What if I need 50 mL of an 8% NaCl solution?

- ✓ Okay. Now that you have your basics down, let's look at the recipe for the hair lysis buffer that you will need to prepare soon. In this case, you need several reagents (the ingredients required for a chemical reaction) mixed together. this means that they will all share the same final volume ($V_2 = 500$ μL) but each one of them will have their own V_1, depending on C_1 and C_2.

✓ How was the 20% SDS stock prepared? Assume 20 mL were made.

Reagents	C1	C2	V1
H₂O	solvent	solvent	
Tris-HCl pH 7.5	1M	10 mM	
EDTA	0.5 M	10 mM	
SDS	20%	2%	
NaCl	2 M	50 mM	
DTT	1 M	80 mM	
Sodium Acetate pH 5.2	3M	240 mM	
PK	25x		
V2			500 µL

Dilutions: the trick

1. Identify C_1, V_1, C_2, and V_2
   - Keep in mind that $C_1 > C_2$ and $V_2 > V_1$. Also, typically you are looking for V_1 (how much you get out of the stock)
2. $C_1V_1 = C_2V_2$
   - Remember that in strength dilutions (10x, 100x, 15x etc...) the "X factor" is C_1/C_2
3. Solve for V_1 — or whatever you are looking for
   - Remember that % dilutions are different! Recall the definition (3% solution means 3g/100 mL) and then solve with a proportion

Make sure to practice dilutions on your own. You have to become confident and independent!

Practice!

1. Watch the units! I have a solution of dNTPs at 10 mM and I want 50 µL of 400 µM dNTPs. What do I mix?

2. I want 1L of working solution from a 120X stock. What do I mix?

3. % dilutions: I need 250 mL of 4.8% cocoa in milk. What do I mix?

Project 1 overview

Analysis of the TAS2R38 locus; Are you a taster?

Objective

The objective of this project is to determine which alleles students carry at the TAS2R38 locus and to verify if the genotype is reflected in the expected phenotype.

This project will span several experiments, as outlined below, over several weeks.

- DNA extraction
- DNA quantification by fluorometry
- Bioinformatics: primer design
- PCR amplification of the Tas2R38 gene
- Agarose gel preparation and gel electrophoresis (to visualize PCR results)
- Amplicon purification and quantification
- DNA sequencing
- Data analysis and Hardy Weinberg law

DNA review

Objective

The objective of these exercises is for you to draw a connection between the chemistry of DNA (its structure) and its function.

Vocabulary

- **bp** - base pairs, this is a unit of measure to define double stranded (ds)DNA length

DNA function

DNA is the blueprint of an organism and is quite stable. Two daughter cells are generated by an identical parent cell during the process of mitosis. During mitosis each cell gets its permanent copy of DNA that will last for the life span of the cell.

Each cell from the same organism has an identical set of DNA sequences. Only part of the DNA is transcribed into RNA. Only a few of the RNAs transcribed are translated into proteins. Within the same organism, cells with different functions have a different set of RNAs, but identical DNAs.

DNA can be used for genetic studies investigating the DNA sequence and the type of genes (alleles) present in an organism.

✓ In 1996 the sheep Dolly was 'cloned' from another sheep[1] through somatic cell nuclear transfer (SCNT). What does that mean (if you don't know, guess, based on the words)?

✓ Why was that such a breakthrough?

DNA structure

Look at the DNA molecule below:

✓ Is it double stranded or single stranded? (1pt)

✓ What's its length in bp (1 pt)?

✓ What molecules constitute the backbone of DNA? Box the sugars in the DNA below and put an arrow by the second molecule, wherever you see it

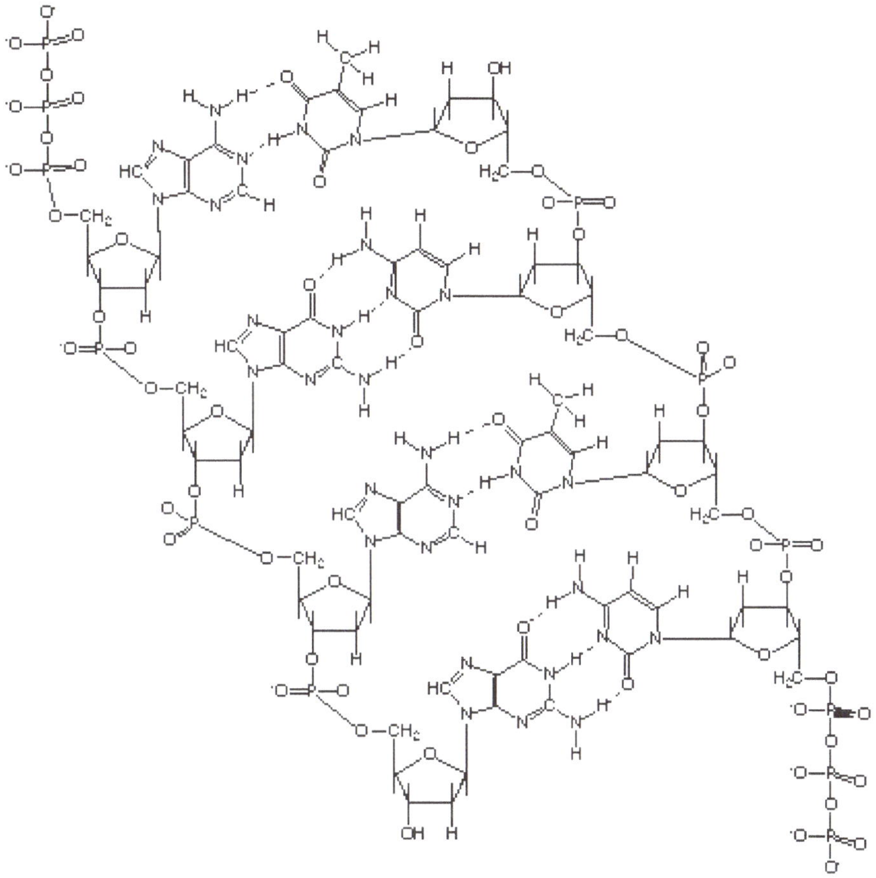

✓ What holds together the two strands? Name it and circle all of them in the figure above

✓ The nitrogenous bases are covalently attached to what?

✓ How many nitrogenous base types can be found in DNA? Write their name below and specify if they are purines or pyrimidines.

✓ Nitrogenous bases are made by rings. How many rings, overall, make the width of a double helix (1 pt)?

✓ What are the two possible pairings? How many hydrogen bonds per each pairing?

✓ Which pairing will be more difficult to disrupt?

✓ Based on the number of rings and hydrogen bonds, identify the bases in the DNA molecule above by writing their initial on the figure

✓ Number the carbons of the four terminal sugars in the figure above. Which DNA end is 3' which 5'? Label the 3' and 5' endings of each DNA strand in the figure

✓ Where could nucleotides be added to the above molecule? Write "ADD" where appropriate

DNA extraction principles

Vocabulary

- **Lysis** - to break down cells
- **Saline** - a solution like PBS (phospate buffer saline) that has the same tonicity of cells (~300 mOsm, isotonic solution)
- **Centrifugation** - using centrifugal force to accelerate sedimentation of a mixture
- **Resuspend** - to put back into solution
- **Pellet** - a small aggregate of particles. In the lab it is generally obtained by centrifugation
- **Decant** - to pour, discard
- **Supernatant** - the liquid above a pellet or a precipitate
- **Buffer** - a solution that minimizes pH change

Overview of the protocol

Steps

In order to extract the DNA we will have to:

1. Collect hair **(Sample collection)**
2. Break the cells' membranes to free the DNA **(Cell lysis)**

3. Separate DNA from the other macromolecules **(DNA extraction proper)**
4. Precipitate the DNA so that it's not too diluted **(DNA precipitation)**
5. Resuspend the DNA in a small volume based on experience **(DNA resuspension)**
6. Quantify the DNA **(DNA quantification)**

Principles

Lysis

How do we lyse the cells?

✓ What are cellular membranes mostly made of (1 pt)?

✓ What do you typically use to get rid of that (1 pt)? HINT: What do you use to wash dirty dishes?

Indeed, SDS is a detergent. You will see; it foams. Also, we'll have to get rid of proteins. To do that we will use an enzyme that destroys proteins, proteinase K, and a strong denaturing chemical, DTT (dithiothreitol).

DNA extraction (separation from other macromolecules)

✓ What are the main four classes of macromolecules (1 pt)?

✓ Is DNA hydrophilic or hydrophobic (1 pt)?

✓ Are lipids hydrophilic or hydrophobic (1 pt)?

✓ Are proteins hydrophilic or hydrophobic (1 pt)?

✓ Which is heavier, phenol or water (1 pt)?

DNA can be easily separated from lipids, based on polarity. Remember? Like dissolves like. Phenol is a lot less polar than water. Therefore, if you mix your sample with a phenol-chloroform solution, hydrophobic molecules will tend to leave water and go into the phenol, forming two distinct phases: an aqueous one with DNA, and a phenol one with fats...and proteins.

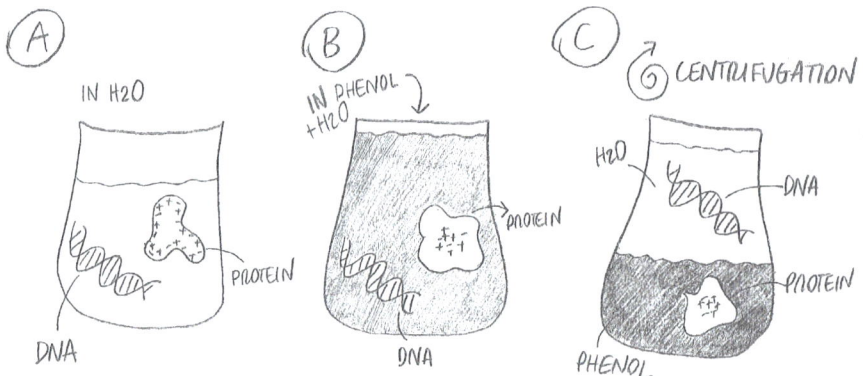

Phenol extraction of DNA. The ambivalent nature of protein's polarity allows for separation of proteins from DNA in an aqueous solution of phenol. While in water proteins will display their hydrophilic amino acids, once mixed with phenol they will denature exhibiting hydrophobic amino acids and becoming therefore nonfunctional yet soluble in phenol.

But why would proteins separate from DNA in phenol? In order to separate DNA from proteins, we are taking advantage of the <u>ambivalent nature of proteins' polarity</u>.

Proteins are made of amino acids, some are polar, some are not. In the cell, proteins that are immersed in an aqueous environment have their tertiary (or quaternary) structure arranged to expose polar residues toward and hide hydrophobic ones toward the inside of the protein (see figure above, left).

When we mix our samples with phenol, proteins turn inside-out with their non-polar residues oriented toward the solvent (figure, middle). Separating phenol from water will drag proteins into the organic phase (phenol) leaving DNA in the aqueous phase (figure, right).

> ✓ So, after mixing your sample with phenol and separating water and phenol by centrifugation, in which phase will lipids go (1 pt)? Proteins (1 pt)? DNA (1 pt)?

DNA precipitation

To precipitate means to disrupt the interactions between a solute and its solvent, so that we can separate them. To precipitate DNA, we need to consider why DNA is water-soluble.

> ✓ What is the basic structure of DNA? Any charges? Polar groups? (1 pt)

So, DNA mixes with water mostly beach the negative charges of the phosphate groups interact with water. Adding a lot of salts (ionic compounds) to a DNA solutions will cause the positive ions to interact with the negative charges on the DNA, effectively displacing water. If a less polar solvent, like ethanol, is added to the samples, it will shield water from interacting with the charges on the DNA. Colder temperature facilitates precipitation because high temperature provides energy, which helps molecules "agitate" and mix with each other.

In short, to precipitate DNA, we will need: 1) high ionic concentration, 2) ethanol, 3) cold. In this way the DNA is basically forced out of the water, forming salts with the ions and precipitating out of solution.

DNA Resuspension

DNA is now a pellet (centrifugation will facilitate precipitation) and can now be mixed with pure water to obtain a pure solution of DNA, whose concentration will be determined via the last step: quantification by spectrophotometry.

Principles of Spectrophotometry

Objective

The objective of this part of the lab is for you to become familiar with spectrophotometry. We will use spectrophotometry to quantify your extracted DNA, but it can be used to quantify DNA, RNA, or proteins with slightly different protocols.

Vocabulary

- **Spectrophotometry** - quantifies the concentration of a sample based on the amount of light traveling through or reflected by the sample at a specific wavelength
- **Sensitivity** - how sensitive is an assay, in other words what is the smallest value it can record
- **Precision** - the reproducibility of an instrument or assay
- **Standard curve** - a series of known samples that will allow you to correlate two parameters; typically concentration and the readout on an instrument

Background

Very often, in molecular biology and genetics you need to quantify very small samples of DNA, RNA, proteins etc... Spectrophotometry measures the amount of light going through a sample or, in other words, the light which, at a specific wavelength, is not 'absorbed' by a sample.

According to Beer-Lambert Law, the more light is absorbed, the higher the concentration of the sample. Each specific assay using spectrophotometry will have a different specificity and precision.

Nucleotides, and therefore DNA, absorb ultraviolet light, with a maximum absorbance at 260 nm. Absorbance is unitless. The know relationship between DNA absorbance and its concentration reveals that **50 µg/mL of pure DNA have an absorbance at 260 nm of 1**.

Using this relationship, you can determine the concentration of your DNA sample.

To determine the **purity** of your sample, the Nanodrop, a spectrophotometer able to work with as little as 1 µL, will measure two absorbance ratios: 260/280 and 260/230.

The **260/280 ratio** should be above **1.8 for pure** DNA. Because both proteins and phenol absorb at 280, a 260/280 ratio lower than 1.8 typically indicates contamination with either.

The **260/230 ratio** should be **2-2.2 for pure DNA**. Because both phenol and carbohydrates absorb at 230, a 260/230 ratio lower than 2 typically indicates contamination with either.

- ✓ Which 260/280 ratio indicates that your DNA is pure (1 pt)?
- ✓ Which ratio might indicate protein contamination (1 pt)?
- ✓ Which ration might indicate phenol contamination (1 pt)?

The Nanodrop

Our spectrophotometer is a Nanodrop. This is a very accurate instrument for the quantification of RNA, DNA, and proteins. The Nanodrop is ecological because it requires no plastic tubes and the procedure is rather simple. Since absorbance depends on the length of the path that the light has to cross through the sample, spectrophotometry normally uses disposable or quartz cuvettes of a specific size. The Nanodrop, ingeniously uses two mechanical arms to squeeze a droplet of sample (1-2 µL) into a fixed width, so that no cuvette is needed.

The instrument needs to be wiped to remove the sample.

- ✓ Why aren't we concerned with sample cross-contamination?

To have an accurate quantification, we will first need to read a droplet of what was used to dissolve the DNA (water in our case) to use as a "blank," so that the Nanodrop will be able to subtract from our sample the background absorbance of the solvent.

The genetics lab 101

Lab rules

- No food or drinks allowed in the lab
- Reduce, reuse, recycle
- Social distancing and mask on at all times
- Do not touch anything you don't know!
- Uncertain? Always ASK
- Always clean after yourself: the threat of unknown spills
- Genetics and molecular biology are HARD. Never let a negative result get you down! Micropipette operation

Exercise: Selecting pipets and setting the volume

Micropipette volumes are set in microliters (μL).

$$1\,\mu L = 10^{-3}\,mL = 10^{-6}\,L$$
$$1\,mL = 1,000\,\mu L = 10^6\,nL = 10^9\,pL$$
$$1\,L = 10^3\,ml = 10^6\,\mu l = 10^9\,nl$$

Each micropipette is intended for use within a volume range. NEVER USE A MICROPIPETTE OUTSIDE ITS RANGE! They lose calibration (become less precise) and might break.

Locate the volume range on the side of your pipettes. If you only see one volume it is the maximum volume. Assume the minimum volume to be 10 times less. Work in pairs

- ✓ Which pipette would you use to pipette 445 μL? Set it
- ✓ 0.5 ml? Set it
- ✓ 18 μL? Set it
- ✓ 1.5 μL? Set it

Exercise: pipette operation

These are the steps involved in pipetting correctly. They will soon become second nature to you.

1. Select a p200
2. Set the volume at 80 μl, as appropriate based on the pipette's range by checking the dial and operating the range wheel. If there is a line separating digits, it separates decimals

3. Select the appropriate disposable tips and press the pipette onto one
4. Remove the pipette with the tip from the box
5. Press the piston: the piston expels the same volume of air as indicated on the range wheel
6. Submerge the tip right below the surface of the liquid you want to draw and SLOWLY release the piston. This creates a partial vacuum and the specified volume of liquid is aspirated into the tip
7. When the desired volume has been drawn, transfer to the first well of your plate leaning against the inside wall, if you had already liquid in the well, you would dispense into it. To dispense, press the piston to the first stop again
8. Repeat for all 8 wells in your row
9. Check at the bottom of your wells. Are the volumes consistent?
10. Pipette the 80 μl back from each well into the original tube without making bubbles. Is there any leftover liquid? Repeat till you build some confidence
11. Set a p1000 to its maximum and try to draw water slowly without any air bubbles

DNA extraction: Sample collection and lysis

Materials

- Lysis buffer will be made of: 10 mM Tris-HCl, 10 mM EDTA, 50 mM NaCl, 2% SDS, pH 7.5, ~80 mM DTT, 240 mM of sodium acetate, pH 5.2, and proteinase K (to a final concentration of ~0.3 mg/ml)
- 65 °C water bath
- Tweezers and scissors
- Table top centrifuge
- 1.5 ml tubes
- 2 ml tubes

- micropipettes
- Disposable tips

Experiment 1 Protocol (DNA extraction, part 1: Cell collection and lysis)

This procedure requires the use of gloves from step 4 on.

1. Assemble the lysis buffer in a 1.5 mL tube, according to the following table.

     Remember to start from the biggest volume (unless otherwise indicated), deposit the liquid at the bottom of the tube, and add following reagents into the liquid. Be mindful of contamination! Ensure your micropipette or tip touch nothing else and don't talk over your samples.

Reagents	C1	C2	V1
H2O	solvent	solvent	277.5 µL
Tris-HCl pH 7.5	1M	10 mM	50 µL
EDTA	0.5 M	10 mM	10 µL
SDS	20%	2%	50 µL
NaCl	2 M	50 mM	12.5 µL
DTT	1 M	80 mM	40 µL
Sodium Acetate pH 5.2	3M	240 mM	40 µL
PK	25x		20 µL
V2			500 µL

2. Pull out three hairs with the root (ouchy!). Cut them a few millimeters from the root and drop the roots into your lysis buffer
3. Vortex and incubate for 2 h at 65 °C
4. (I will do this). Vortex, add 40 µL 1M DTT and 15 µL Proteinase K (10 mg/ml)
5. Mix gently and incubate 2 h at 60 °C
6. Store at -20 °C till the next time

DNA extraction (proper) and precipitation

Experiment 1 Protocol (continued, DNA extraction, Part 2: extraction and precipitation)

Remember? Last week we lysed the cells and I put the samples in the freezer. Because this protocol could be done all in one day, you'll notice the numbers continue from last time.

7. Retrieve your sample from the freezer.

- ✓ What's in your tube now (1 pt)?

Now that we lysed the cells, we need to separate the DNA from the other macromolecules by polarity by adding phenol (DNA extraction), precipitate the DNA, resuspend it, and quantify it.

8. Separate the DNA from proteins by adding 1 volume (500 µl) of Phenol: chloroform: isoamyl alcohol (PCI) = 25:24:1. CAP SECURELY and shake 30" at room temperature.

CAREFUL! Phenol is toxic! Wear protective glasses, roll down the sleeves of your lab coat. Shake the tubes away from your body and over the bench (not over the manual). If there is a spill, call me right away and I will clean it. If you soil your gloves replace them.

- ✓ Why are we shaking the tubes?

9. Centrifuge for 7 minutes at 11,000 g to separate organic and aqueous phase

- ✓ What will cause separation in two phases (1 pt)?

Centrifuges can be operated in **rpm** (revolutions per minute) or in **g** (acceleration). The acceleration is a more accurate unit because it indicates the actual force applied to your sample. Conversely, rpm indicate speed. Force will depend on both speed and the radius of the centrifuge utilized.

- ✓ Is it more accurate to write a protocol specifying the speed of centrifugation in g or rpm (rotations per minute)? Why (1 pt)?

- ✓ In which phase will your DNA be (2 pt)?

10. Label a clean 1.5 mL tube with your lab number
11. Transfer the aqueous phase (upper phase) in the clean tube with a p200 set at ~180 µL. Avoid the white matter at the interphase, which mostly contains denatured proteins
12. Repeat steps 8, 9 and 10 to increase DNA purity and transfer into a final clean rube labelled with your number and DNA (e.g. 12 DNA)
13. Precipitate DNA with 1/10 of volume (50 µl) of 3M Sodium Acetate pH 5.6 and 2 volumes (1 ml) of cold absolute ethanol. Invert the tube several times
    - ✓ Can you see the DNA precipitate? (It looks like a small white wispy cloud). Mark your answer here and in the results section at the end of the manual

If yes, you will likely have a good yield, YAY! If not, don't despair, you still might have some DNA.

14. To facilitate precipitation incubate your sample at – 20 °C. This incubation could be as short as 15 minutes, but we will finish our protocol next week

DNA extraction: Resuspension and Quantification

Experiment 1 Protocol (continued, DNA extraction, Part 3: resuspension and quantification)

Remember? Last time we extracted and precipitated your DNA.

15. Retrieve your sample from the -20 °C freezer and centrifuge 15 min at 4 °C at 11,000 g
    - ✓ Where will your DNA be at this stage, in the pellet or the supernatant (1 pt)?

16. Visualize your pellet. Ask for help if you can't see it

17. Carefully decant the ethanol in a sink without disturbing the pellet. Use a p200 and a p10 to remove any traces of ethanol. Ask me to check your sample before moving to the next step, as traces of ethanol might inhibit later experiments

18. Resuspend in 50 μl of molecular biology grade water. Pipette gently your sample up and down to minimize dispersion (spreading your sample all over the tube is NOT good).

    This sample should contain your own DNA. However, 1) We don't know if it does and 2) If it does, we don't know how much, which would make it harder to use for later experiments. For this reason we will use spectrophotometry to quantify the amount of DNA, if any, in your sample.

Experiment 2: spectrophotometer quantification

Materials
- Extracted DNA
- Molecular grade H2O (H tube)
- Nanodrop spectrophotometer
- Micropipettes

- Filter tips

Protocol

 We will carry out the quantification on a small portion of your sample (2 μL), which will be then thrown away. Nonetheless, the quantification will tell us the concentration of the remaining sample (48 μL), which you will keep for further experiments! (In our case PCR and sequencing analysis).

1. Watch me load the blank on the pedestal of the Nanodrop
2. When your turn comes, load 2 μL of your sample onto the pedestal. Take care not to scratch the lens.
3. Gently close the upper arm and, with my help, read your sample. Write your data below and in the Results section:
   - ✓ My DNA concentration is (record the unit!):
   - ✓ My 260/280 ratio is:
   - ✓ Therefore I conclude:
   - ✓ My 260/230 ratio is:
   - ✓ Therefore I conclude:
   - ✓ The total volume of my sample (including what was used for the quantification) was:
4. Sometimes in science it is useful to calculate the total yield; the total amount of what was extracted. The total yield indicates how well your protocol worked and of course will depend on the size of the initial sample. The total yield is the total amount of a solute in a sample and is calculate multiplying the concentration of the sample by its volume. In this experiment, the solute is DNA.

$$\text{Tot yield} = C \ast V$$

- ✓ What was your total yield of DNA (with appropriate units)?
- ✓ Did you obtain any DNA?

- ✓ If not, hypothesize what might have gone wrong in your procedure

- ✓ Go to the results section at the end of the manual and fill
- ✓ For example, imagine you extracted RNA. The concentration of your sample is 5 mg/mL. Your sample is 300 µL. What's the total yield?

PCR theory

We want to study one specific gene, but every cell has only two copies of each gene and they are so very tiny! To be able to get enough copies of a specific gene we will have to become very ingenious. That's what Kary Mullis did when he invented the Polymerase Chain Reaction in 1985 (PCR, Nobel prize 1993) and became the father of modern molecular biology!

Objective

The objective of this lab is to understand the principles of PCR.

Vocabulary

- **Primer** - a short, typically 20 bases, stretch of DNA complementary to a region of nucleic acid
- **Primer set** - Two primers designed to work together in a PCR experiment
- **Annealing** - forming a double stranded nucleic acid by complementarity by creating the appropriate hydrogen bonds
- **PCR** - polymerase chain reaction, is an experimental technique used to obtain many copies of one specific DNA region, so that it can be further analyzed or visualized
- **Annealing temperature (Ta)** - the temperature at which primers pair with their complementary DNA
- **Melting temperature (Tm)** - the temperature at which 50 % of the primers are still attached to their complementary DNA, a few degrees above the Ta
- **Amplicon** - the region of DNA included within the primer set and 'amplified' (copied over and over) by PCR
- **SNP** - Single Nucleotide Polymorphism. A point mutation present in part of the population either because selected or because associated with a nearby sequence that was selected. It may or may not be associated with a phenotype
- **Thermal cycler** - A machine that quickly and precisely ramps up and down to programmed temperatures
- **Taq** - the DNA polymerase from Thermus Aquaticus

Background

Alleles often differ only at one or few bases. Single Nucleotide Polymorphisms (SNP) are differences of one base that in some cases are enough to generate a different phenotype. If the phenotype is favorable, this random mutations might be selected through thousand of generations and a new allele arises.

An allele at the TAS2R38 gene gives people the ability to taste certain bitter substances[3], possibly giving them an evolutionary advantage as certain poisons are bitter. The "taster" allele is different from the wild type by 3 SNPs.

Are you a taster?

You extracted your own DNA in the last weeks and you would love to check if you carry the taster allele, but how do you find and isolate that specific sequence from all of your DNA?

Kary Mullis is a biochemist. In the 80s he was pondering about the very same question.

Like you, he knew that DNA polymerase duplicates DNA during cell division.

He wondered if he could trick DNA polymerase into making many copies of only the gene that he wanted to study. Eventually he managed, and invented the polymerase chain reaction (PCR), winning the nobel prize and the title of father of modern molecular biology[4].

To understand how PCR works we need to review a couple of concepts...

- ✓ What allows DNA polymerase to start its 'copying' activity on single stranded DNA (1 pt)?

So, Kary Mullis had the idea to design a synthetic primer, complementary to his gene of interest.

- ✓ If you were to mix DNA polymerase and your DNA in a test tube to get a copy of your gene TAS2R38, what other ingredients would you need?
    - ✓ A bivalent ion (X^{2+}); a required cofactor for DNA polymerase to work
    - ✓
    - ✓

- ✓ DNA is complementary and antiparallel. Given the DNA strand below, draw its complementary strand.

5'-TGTAA**ATG**CCCGTATGTAGGCATGGCAT..TTAGGAT-3'

- ✓ Why, do you think, is ATG in bold?

- ✓ Which one is the coding strand (1 pt)? Which one is the template strand? Write it above

- ✓ If DNA polymerase were to synthesize the strand you just drew (provided it had a primer), would it proceed right to left or left to right? Why (2 pts)?

> To summarize: DNA polymerase needs dNTPs, bivalent ions, a _____ and a template to start synthesizing DNA in direction _____' to _____'.

- ✓ If, theoretically, you were trying to get many copies of ONLY the coding strand printed above, not the one you wrote in, would your primer be complementary to the coding strand or the template strand (1 pt)?

- ✓ Would your primer be identical to the coding strand or the template strand (1 pt)?

- ✓ Would your primer pair to the left or to the right of its complementary strand (1 pt)?

As it happens, having only one primer would not work at all. Let's see why. (Is your brain exploding? That's okay! This stuff is complicated!)

- ✓ Assume you were starting from one human cell. How many copies of the TAS2R38 gene would be available for DNA polymerase to copy?

- ✓ After **one cycle of 'amplification'**, meaning that your primer pairs to its complementary strand and DNA polymerase fills in a copy of the gene, how many copies of TAS2R38 would you have in your tube (1 pt)?

- ✓ And after another cycle of amplification?

> An efficient PCR experiment uses a **primer pair**: two separate primers of which the first one is complementary to the coding strand, the second one is complementary to the template.
>
> The primer that is identical to the coding strand (complementary to the template) is called **sense** or **forward primer**.
>
> The primer that is identical to the template strand (complementary to the coding) is called **antisense** or **reverse primer**. This primer will be complementary and reversed to the coding strand.

The figure below shows what would happen if we were doing PCR starting from just 2 copies, like in the previous example, but with a primer set. Remember that when you extract DNA you start from many cells, typically several millions! So you would have a lot more than two copies.

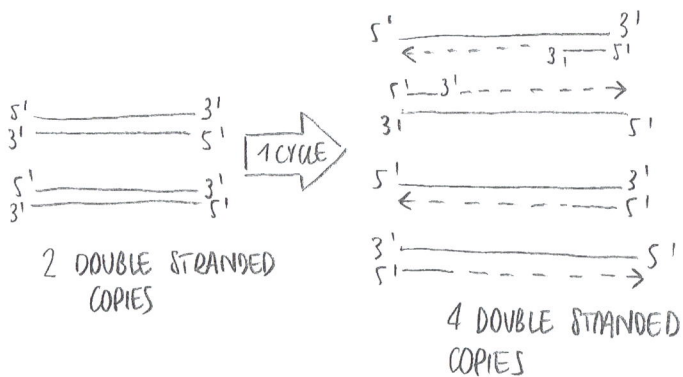

PCR amplification of DNA after 1 cycle (starting from the DNA of one diploid cell)

- ✓ How many double stranded copies would you have after the first PCR cycle with a sense and antisense primer (1 pt)? HINT: look at the figure above

- ✓ How many double stranded copies would you have after a second PCR cycle (1 pt)?

- ✓ How many double stranded copies would you have after a third PCR cycle? (1 pt)

- ✓ How many double stranded copies would you have after a fourth PCR cycle?

- ✓ If you started from 10 copies of dsDNA, how many copies would you have after 1 cycle? After two cycles? (1 pt each)
- ✓ What's the generalized formula to find how many copies you would have after n cycles of PCR if you started from "a" number of copies (5 pts)?

PCRs typically run for 25-35 cycles, that's a lot of amplification!

- ✓ If you had 2,000 copies of your gene and ran a PCR for 35 cycles, how many copies of the gene TAS2R38 would you end up with (2 pts)?

- ✓ If you had DNA from 2×10^6 cells and ran a PCR for 35 cycles, how many copies of the gene TAS2R38 would you end up with (4 pts)?

PCR set up

So let's see how PCR works, step by step. As you will see, temperature control is essential. For ease of understanding, let's just work with one double stranded DNA containing your gene of interest (below).

A simplified sketch of an example gene (One gene= one ds DNA)

✓ Could our primers anneal if the DNA were double stranded? Why (1 pt)?

✓ The first thing we need to do then, is to separate the two complementary strands, so that we can copy them. How could we do that (1 pt)?

✓ Next, we need our primers 'to anneal'. How do we do that (1 pt)?

> The temperature at which half of the primer pairs to its complementary DNA is the **melting temperature** (**Tm**). The Tm depends on the length of the primer and its base composition. A rule of thumb to determine Tm for primers is:
>
> $$Tm = 2\,°C \times (\#AT) + 4\,°C\,(\#CG)$$
>
> Where #AT is the the total number of As and Ts in the primer, and #CG is the total number of Cs and Gs. Cs and Gs require more energy (heat) to separate because they are held together by 3 hydrogen bonds, As and Ts only by two.

✓ Now we need DNA polymerase to do its work. At what temperature, do you think, DNA polymerase will work (1 pt)?

✓ You are through with one cycle. Although doable, Kary Mullis's PCR protocol was far from easy. Dr Mullis had to run his tubes from one water bath to the next, set at different temperatures, for the whole length of the experiment (several hours). Also, what do you think happened to DNA polymerase every time he put his tubes into a 95 °C water bath (1 pt)?

So, he had to add new polymerase at each cycle, which made the experiment very expensive and labor intensive.

It was only in 1988 that Saiki and his coworkers thought to employ in PCR a heat resistant DNA polymerase isolated from Thermus Aquaticus, a bacteria thriving in geothermic active areas[5]. This enzyme, named Taq, is nowadays used for PCR experiments all around the globe.

The addition of thermal cyclers, machines that can be programmed to quickly cycle through temperatures, made PCR experiments easier, but it is still up to you to avoid pipetting errors and contamination!

✓ Okay! So, let's sketch what a PCR "program" would look like (steps, temperature, durations, meaning)

General DNA PCR guidelines
PCR reactions are typically carried out in a 15-50 μl volume containing 1 ng-1 μg of genomic DNA (100 ng is ideal) in buffer, primers (100-500 nM), dNTPs (0.2 mM), $MgCl_2$ (0.5-4 mM), and Taq (amount recommended by the manufacturer, based on the enzyme's activity).

When working with PCR always use gloves and filter tips!

✓ Why should you always wear gloves?

✓ What, do you think, is the function of the buffer (1 pt)?

✓ And MgCl2 (2 pts)?

✓ Do you expect your DNA to be positive or negative for the TAS2R38 gene? Why?

Set up

This protocol is based on the AmpliTaq Gold 360 kit by Thermo Fisher Scientific. Always check the specific protocol for the kit you are using and adjust as needed.

We want to add to the reaction 10 μL (it fits and it's more than 5 μL).

✓ Given your DNA concentration (see the results) how much DNA would be in 10 μl? Show calculations below.

✓ Is that amount within the allowed range (1 ng- 1 μg)? If not, ask for my help.
✓ Fill the table below:

Order and ✔	Reagents	C1	C2	V1
	H2O	solvent	solvent	
	primers	4 μM	200 nM	
	Buffer	10X		
	MgCl2	25 mM	1.4 mM	
	dNTPs	10 mM	0.2 mM	
	TAQ			0.6 μL
	DNA			10 μL
	V2			50 μL

- ✓ To avoid mistakes and pipetting small volumes, we often resolve to use a "**master mix**", a mix that contains all the reagents common to all the samples, including positive and negative control when appropriate.

- ✓ What is a master mix and why is it useful?

- ✓ What would have a master mix looked like for your DNA extraction?

- ✓ Let's workout our master mix recipe, which is what you will do once in the lab. Working as a (socially distanced) group, we will have a master mix for six students +4 (negative control, positive control, and two extra samples to account for dead volumes).

Order and ✔	Reagents	1x	10x
	H2O	28.1 µL	
	primers	2.5 µL	
	Buffer	5 µL	
	MgCl2	2.8 µL	
	dNTPs	1 µL	
	TAQ	0.6 µL	
	DNA	10 µL	NO!
	V2	50 µL	

- ✓ Why is there no DNA in the master mix? (1 pt)
- ✓ Calculate the total volume in the 10X master mix and divide by the number of samples. Does the result make sense? Why?

PCR Tips

- Make calculations for your master mix first, and double check them. THEN, each student takes turns to add the ingredients in the correct order <u>and adds a check mark after adding each one of them by the table</u>

- CHECK ON EACH OTHER! If one student makes a mistake he's going to screw up everyone's experiment. <u>Learn to work as part of a team</u>

- Keep all ingredients on ice to minimize degradation. Taq degrades easily. For this reason I will hand out Taq aliquots only when you need it. Be quick to add it and return it to me, but make sure that you have some Taq in your tip

- Always calculate the total of μl and write it in the table, divide the total by the number of samples you prepared: how many μl per sample? Is it correct? In which order should you add the reagents? Write it in the table below (left column)

The importance of controls

Objective

The objective of this part of the lab is for you to understand the concept of reproducibility, error and controls.

Background

Now that you know a bit more about molecular biology, you can appreciate how easy it is to make a mistake that you might not be able to detect. You might make an error pipetting, some reagent might be deteriorated or you could introduce contaminants. How do you "control" for these factors?

Vocabulary

- **Precision** - how close to the true value the measured value is (how could an instrument be precise but not accurate?)

- **Sensitivity** - what's the smallest value the instrument can measure?

- **Negative control** - a sample that will be negative in your experimental setting unless contaminated

- **Positive control** - a sample that will be positive in your experimental setting unless a technical mistake occurred
- **Technical replicates** - the same sample ran multiple times within the same experiments
- **Biological replicates** - repeating an experiment starting from different biological samples
- **Reagents** - the ingredients of an experiment

Positive and negative controls, technical and biological replicates

To understand which controls you need, every time you are setting up an experiment imagine the possible outcomes. For example:

- ✓ Do you remember any "controls" during our DNA extraction?

- ✓ What were the possible outcomes?

- ✓ What would you have concluded in either case (2 pts)?

- ✓ Do you expect your DNA to be positive or negative for TAS2R38? Why?

- ✓ You run your PCR and your sample is positive for TAS2R38. How can you rule out that your sample is contaminated?

- ✓ Your sample is negative. What would be your possible explanations? How could you tell them apart?

To formalize the importance and significance of positive and negative controls, let's look at some hypothetical scenarios. You set up a PCR for TAS2R38 with a group of 4 students and:

A- All samples are positive. Terrific! How can we be sure that we did not contaminate all of our samples, which would have otherwise been negative? You need a **negative control**, a sample that is expected to be negative unless contaminated

B- Everything turns out negative. Bummer! But how do we know that our samples are really negative and that we didn't make a mistake in setting up our experiment? Our samples might look negative because Taq did not work. We need a **positive control**, a sample that we know to be positive in our experimental conditions

C- We had our positive and negative controls, so our data is golden. But what if something happened only to a specific sample (e.g. I sneezed on it, I forgot to add the final reagent, etc...)? We need **technical replicates**, meaning that you will run within the same experiment the same sample two or three times. This is done only in some experiments (Typically not traditional PCR)

D- OK! We had all of the above, are my results good, yet? Well, now we need to prove that whatever we observed is not just a fluke of nature, but happens consistently. We need **biological replicates** (usually triplicates). This means repeating the same experiment from different biological samples several times. We will not do this in our lab, but it need to be done for any actual research before publication

Now that you understand the principles of PCR and the importance of controls, we can focus on primer design.

DNA Primer design: genetics meets bioinformatics

Today you will have a sneak peek into bioinformatics: computer science applied to Biology.

The human genome has 3×10^9 base pairs including about 22,000 protein encoding genes, transcribed into 140,000 isoforms [2].

So, how do we find, study, analyze, and sequence one specific portion of the genome?

The answer lies within bioinformatics and primer design...

Objective

The objective of this lab is to learn to design primers and become acquainted with bioinformatics. The next time, we will use the primers we design today to run a TAS2R38 PCR on your DNA.

Vocabulary

- **Bioinformatics** - computer science applied to biology, generally to browse and analyze large sets of information or data

Experiment 3: Bioinformatics, DNA primer design

Materials

- A computer

Take copious amounts of notes, as primer design makes for an ideal assignment.

Overview

In order to design specific, effective primers we will have to:

- Look into databases to retrieve the DNA sequence of the gene we are interested in
- Select the regions where we want to design our primers (this will change each time, based on what you need to study! In this case we want to design primers around the SNP that is different between the taster and non taster allele.)
- Design the primers, so that they are specific (amplify only the gene of interest) and effective (amplify a lot of it). We will learn later in the semester how to check for specificity

We are going to design a primer set to amplify the gene TAS2R38. We want our **amplicon** (the region copied by PCR) to be about in between 300-500 bp, and to include the region of the SNP that allows some people to taste glucosinolates.

The protein encoded by tasters has a proline instead of an alanine at position 49 (A49P change).

✓ Looking at the genetic code in the figure below, which single base change could have caused an alanine to become a proline (1 pt)?

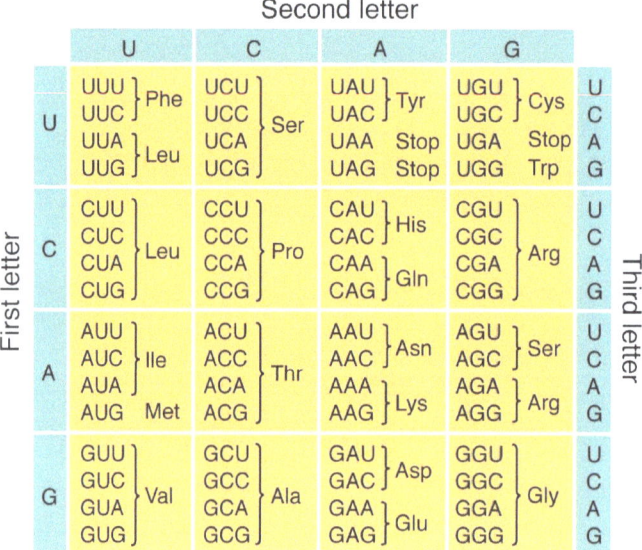

The genetic code

We will have to look at the DNA sequence to find out which one is the case!

Protocol

Finding the correct gene

There are many databases available for free online. A couple of the most popular ones are:

-The National Center for Biotechnology Information (www.ncbi.nlm.nih.com)

-Ensembl genome browser (www.ensembl.org)

Although these genome browsers have a lot of information, they are sometimes hard to navigate, especially for a beginner.

Many gene sequences might come up after your search; which one is the right one?

For this reason I recommend using a browser from the Weizmann Institute of Science, gene cards: "a searchable, integrated database of human genes that provides comprehensive, updated, and user-friendly information on all known and predicted human genes" (www.genecards.org).

Here you will find much of the information you need for your primer design assignment.

1. Google gene cards and click on www.genecards.org
2. On gene cards search for TAS2R38 and click on the correct gene card (when you do your own assignment, you will click the die to get a random gene)

Here you will find background, genomic location, size of the gene (under genomic views), size of the protein (under proteins) and transcript size (under transcripts). Beware of plagiarism! Reword the information in your assignments!

3. To open the gene sequence under genomic views, click the first RefSeq DNA sequence (NC_000007.14). RefSeq is a repository of verified sequences from NCBI. (When working on your random gene for the assignment, if there are multiple sequences available choose the first one. Never choose a sequence starting with XM_ or XR_)

    ✓ XM_ and XR_ sequences stand for...

The NCBI sequence opens up. Now we are speaking bioinformatics! It looks complicated, but it's really not too bad! There are three sections to this page:

- Comments: any annotation, comment and reference to the sequence
- Features: very informative. It tells you the length of both gene and mRNA, and most importantly where the coding sequence (CDS) starts and ends. The CDS is the part of the sequence that—once transcribed into mRNA—will be translated into a protein

    ✓ Which one do you expect to be longer, the CDS or the gene sequence? Why (3 pt)?

- Sequences: the amino acidic sequence of the protein and the DNA sequence of the gene

Notice that you can skip to any of these three sections from the 'go to' menu at the top of the page.

Moving around the gene

It might seem a bit disorienting at first, but it is really not too hard to move around a sequence. Let's practice a bit.

- ✓ What is a CDS (1 pt)?

If beside the CDS you read "complement", it means that the coding sequence is on the complementary strand.

- ✓ Are we looking at the template strand or the coding strand? (1 pt)

Always work on the coding strand! To get the coding (rather than its complement) go on the top right corner on the window and check "show reverse complement". Then click "update view"

- ✓ Where does the CDS for TAS2R38 start (3 pt)?

- ✓ What amino acid will you expect to find there (1 pt)?

- ✓ Is that indeed the first amino acid of the protein?

- ✓ What codon encodes for it (1 pt)? If you don't remember, look at the genetic code on the previous page

Let me show you how to easily locate that codon into the sequence...

By chance, TAS2R38 has only one exon. Normally, when looking at a CDS, you will see something like this:

```
CDS             join(1161..1190,2815..3586,4963..5525)
```

- ✓ What does it mean? (5 pts)

✓ We said before that the difference is at amino acid 49; tasters have proline (P) and non tasters have alanine (A). Look at the amino acid sequence. Was the person whose DNA was sequenced a taster (1 pt)?

We had four potential codons that could have turned from alanine into proline. Let's see which one we are dealing with.

CHALLENGE!

> The mutated amino acid is A49P...for FIVE king/queen points, what number in the DNA sequence will identify the nucleotide at the beginning of the codon?

✓ What was the alanine codon in this sequence?

✓ What change could turn it into proline?

✓ If we want to design primers around the SNP, so that the region with the SNP will be included in our amplicon, where should we search for the forward primer?

✓ And the reverse? (Refer to the window in the right uppermost corner of the NCBI sequence. Those numbers indicate where your sequence starts and end on the chromosome.

Keep in mind that while your sequence is shown from nucleotides 1-1143, when you tell the computer to search for primers, you have to transpose numbers to the chromosome. Basically, Tas2R38 does not start at nucleotide 1 of chromosome 7, it starts at nucleotide 141,972,631-141,973,773

O_o

✓ For the F primer, the range where to look for the primer will be until 10 bases before the SNP, starting 500 bp before (or at the beginning of the gene, whichever results in a shorter range). For the R primer, we are going to calculate the base number from 10 bases after the SNP until the end of the sequence or 500 bp downstream from the beginning of the range, whichever comes first.).Write your ranges below:

TAS2R38 F range (where I look for the forward primer):

TAS2R38 R range (where I look for the reverse primer):

Designing primers

Now we have the sequence and the SNP that we want to include; we can start to design primers.

You could design primers by hand, keeping in mind the following general rules of thumb:

- Primers are typically 20 bp long. This length maximizes specificity
- Primers should have a content of GC around 50%, meaning 50% of the bases should be Gs and Cs, the other half As and Ts. This will keep the Tm around 60 °C
- Primers should have a compatible Ta/Tm
    - ✓ Why (1 pt)?

- The 3' region of the primer is the one that more strongly determines specificity. Why (1 pt)?

- Avoid stretches of the same base (e.g. AAAA).
    - ✓ Why?

- Avoid regions of self-complementarity (they form **hairpins**) or complementarity (they form **primer dimers**) with the other primer
    - ✓ Why?

That's a lot to keep in mind. Fortunately bioinformatics helps.

4. Right click "pick primers" and open it in a new window
5. In the up right corner of the page, under the PCR template/range, you have to tell Primer-BLAST where you wish your primers to be designed. We need the forward to be on the left of the SNP and the reverse on the right. Based on what we discovered so far about the sequence, change the numbers accordingly.
6. Under the Primer Parameter box, select your desired PCR product size (amplicon size) and type "1" as number of primers (primer pairs to be more correct) to return
7. Skip the Exon/intron selection box (we will work with this later in the semester)
8. Under the Primer Pair Specificity Checking Parameters box, select "Genome (reference assembly from selected organisms)" as your database. Then, checkmark "Exclude predicted Refseq transcripts (accession with XM, XR prefix)
    - ✓ What are XM and XR prefix sequences? (2 pts)
9. Before you click "Pick primers" checkmark "Show results in a new window"
10. Click on "Pick primers"
    - ✓ Based on the start and stop of the primers, let's locate them on the original sequence. Is the forward primer identical to the sequence?

- ✓ What about the reverse primer?

- ✓ Why?

> REMEMBER that primer sequences are ALWAYS given 5' to 3'.

11. Name your primers sensibly (to indicate what they amplify and in what organism, as in the examples below) and always write them 5' to 3'. For example the set we just designed will be h-TAS2R38-DNA-F and h-TAS2R38-DNA-R where h stands for human

12. Write the primers' name and sequence into your results section

Nowadays, once properly designed, primers can be ordered from specialized companies that synthesize them for as little as $0.5 per base.

Next time we will carry out a PCR experiment on your DNA to amplify the TAS2R38 gene.

Polymerase chain reaction (PCR)

Experiment 4: DNA PCR for the TAS2R38 gene

Materials

- ice bucket
- 0.2 ml PCR tubes
- Molecular grade H2O (tube H)
- Genomic DNA (that you previously extracted)
- 4 μM each, primers specific for TAS2R38 (mixed together, P tube)
- 10x PCR buffer and Taq from Life Technology (tube B)
- 25 mM MgCl2 (tube M)
- 10 mM each dNTPs (tube D)
- Taq polymerase (T tube)
- Thermal cycler
- Micropipettes and filter tips

Protocol

1. Assemble your master mix in a 0.5 mL tube (one reagent per student) per your calculations below:

Order and ✔	Reagents	1x	10x
	H2O	28.1 μL	281 μL
	primers	2.5 μL	25 μL
	Buffer	5 μL	50 μL
	MgCl2	2.8 μL	28 μL
	dNTPs	1 μL	10 μL
	TAQ	0.6 μL	6 μL
	V2	50 μL	400 μL

2. Close, vortex briefly, and add 40 μl of master mix to your personal PCR tube (0.2 ml), label the <u>top and side</u> of the tube with your lab number and P (for PCR, e.g. 14P). Pass the master mix to the next student.

3. Add 10 μl of DNA to your PCR sample. Pipet up and down a couple of times. Store on ice.

4. As a group, prepare two extra PCR tubes with 40 μl of mix. You will add 10 μl of water to one of these samples (your negative control). Label it - (minus). Do you expect amplification in this sample (1 pt)?

    ✓ What's the point of this sample (1 pt)?

5. Add 10 μl of positive control DNA to the plus tube. The +DNA is DNA that previously worked in the same reaction

    ✓ Do you expect amplification in this sample (1 pt)?

    ✓ What's the point of this sample (1 pt)?

6. Vortex the samples and spin them down, keep samples on ice until ready

7. When everyone is ready, tubes go in the thermal cycler
8. We will start the program when all the tubes are situated

The PCR will take a couple of hours to run.

When the protocol is done I will transfer your samples to the -20°C freezer, till next week

Gel electrophoresis theory

So far, you extracted your own DNA, quantified it by spectrophotometry, learned to design specific DNA primers, and amplified by PCR the gene we wanted to analyze, TAS2R38.

Now we need to see if the PCR worked. In order to do so we will learn about gel electrophoresis.

Objective

The objective of this lab is to understand the principles of gel electrophoresis.

Vocabulary

- **Gel electrophoresis** - a technique used for the separation and visualization of nucleic acids based on size (typical ideal resolution from 100-2,000 bp but up until 30,000 bp)
- **Agarose** - a natural carbohydrate that, when dissolved through heating in a solution, can solidify at room temperature in a gel
- **DNA ladder** - ladders are collections of DNA fragments of known sizes that can be separated by gel electrophoresis providing a term of comparison to estimate size of sample DNAs
- **Wells** - the holes in the gel that allow you to load samples
- **Loading buffer** - a buffer that allows to load the samples and estimate the "front" of the run (how far the smallest fragments migrated into the gel during the run)

Background

✓ Do you recall the size of the human genome? What is it?

✓ How many DNA bases are in the nucleus of a human, diploid cell (1 pt)?

✓ The distance between two steps in the DNA helix is 0.34 nm. If all the DNA in one cell were to be set in one long chain, how long would it be (show calculations for 3 pts)?

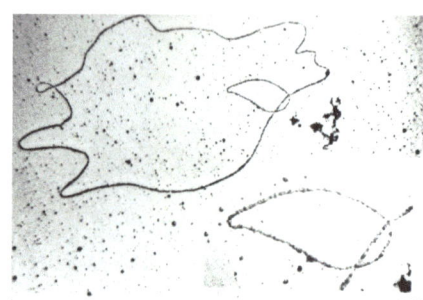 We need microscopes to see cells, can you imagine how hard it would be to visualize DNA!
With electron microscopes (up to 107x!) (see the image aside from the public domain) it is possible to visualize single molecules of DNA, but not in such a detail to be able to tell nucleotides apart.

Other techniques have to be used to "see" a molecule of DNA. Today we will use gel electrophoresis to see if we have a lot of DNA in the size of our expected amplicon (if yes, our PCR worked!). Next, we will use DNA sequencing to "read" the sequence of your alleles, base by base!

Gel electrophoresis

Gel electrophoresis is an easy and affordable technique to visualize large amounts of nucleic acids based on their size. You cannot see "the sequence" of the DNA, just if you have a lot of a fragment of a certain size.

- ✓ For example, if your PCR experiment from last week worked, what do you expect will be in your tube (1 pt)?

Gel electrophoresis consists in sifting your DNA through a mesh.

The smaller is a DNA fragment, the less impeded it will be in proceeding through the mesh (it will go farther from the wells). Large DNA fragments will take longer to disentangle from the mesh and cross it (they will stay closer to the wells).

The mesh is typically made by a gel of either **agarose** or **polyacrylamide**.

Agarose is a natural carbohydrate and it is used to separate nucleic acids 100 bases long and above.

Polyacrylamide, typically cross-linked with bis-acrylamide, can be used to separate a wide range of DNA sizes, including very small ones. Nonetheless it is toxic and more difficult to handle. For this reason agarose is the preferred option, when the application permits it. To put the DNA into the mesh, we will make "wells" at one end of the gel, in which we will "load" our samples.

Different percentages of agarose will allow for separation of DNA fragments of different size, as shown in the table below.

% Agarose	Optimum Resolution for Linear DNA
0.5	1,000–30,000bp
0.7	800–12,000bp
1	500–10,000bp
1.2	400–7,000bp
1.5	200–3,000bp
2	100–2,000bp

✓ Given that the TAS2R38 primers we designed will amplify an amplicon of ~460 bp, what percentage of agarose will you need?

✓ Would it be okay to let the gel boil into the microwave to make sure the agarose dissolve? Why (1 pt)?

To have DNA move through the mesh, an electric current needs to be applied.

An electric current is a flow of charged particles. In an electrophoretic chamber (figure below) a current, supplied by a power pack, flows through the two electrodes (- typically black, and + typically red).

The bottom part of an electrophoresis cell. The gel would be sitting in the gel tray. The chamber is filled with the appropriate buffer and samples are loaded into the gel. The lid is then applied (right) and the electrodes are connected to the power pack, which supplies electricity.

✓ Think about the molecular structure of DNA. In an aqueous solution is DNA charged - or + (1 pt)?

✓ Will DNA migrate toward the + or - electrode?

In order for the current to flow through the agarose, the gel needs to be prepared in a solution (buffer) containing ions, which will allow for the current to flow.

This solution is typically either TAE 1X (Tris-Acetate EDTA buffer) or TBE 1X (Tris-Borate EDTA buffer), pH 8.

The figure below shows how 9 different fragments of DNA would look like, after separation in a gel electrophoresis experiment.

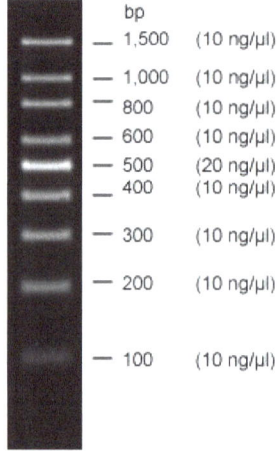

2.0% agarose gel

The 100 bp DNA ladder. The ladder was separated on a 2% agarose gel. The size of each band is indicated.

✓ Look at the figure above and draw the location of the well. Was it at the top or the bottom? (2 pts)

✓ If you loaded the sample into the well, was the negative electrode applied on the top of the gel or the bottom (2 pts)?

✓ Which band of DNA represents a longer fragment, the highest or the lowest (1 pt)?

✓ Can you tell the sequence of the DNA fragments by looking at the gel (1 pt)?

✓ Two fragments are 345 bp long, but have different sequences. Could you separate them by gel electrophoresis (1 pt)?

DNA, per se, would not be visible in a gel. In order to make DNA visible (like in the figure above) and assess the bands, we need to use a fluorescent molecule that binds DNA. This molecule will make DNA visible when exposed to UV light. The fluorescent molecule we use is non toxic (unlike Ethidium Bromide, which is carcinogenic and commonly used in gel electrophoresis).

✓ Is UV light harmful (1 pt)?

Because the fluorescent molecule is not visible under normal light and we cannot run the gel under UV light (it degrades DNA pretty quickly!), we need a dye that migrates similarly to the smallest fragments of DNA to tell us how far the DNA migrated into the gel, so that we can stop the run before the smaller DNA fragments exit at the other end of the gel.

✓ What would the problem be, if we did not have such a dye (1 pt)?

For this reason samples are mixed with a DNA loading buffer before being loaded in a gel.

DNA loading buffer typically contains:

- Glycerol, to make the sample heavy enough to sink at the bottom of the well

- A dye. For example bromophenol blue, which migrates like a 400-500 bp fragment of DNA in an agarose gel 0.5-1.5% and like a 50-150 bp DNA fragment in a 2% gel

- Our loading buffer also contains a fluorescent molecule that sticks to DNA, allowing us to visualize it later under a UV light

✓ Based on what you learned so far about gel electrophoresis, draw what you would expect your PCR sample to look like after separation in the gel electrophoresis chamber. Remember that your amplicon was 460 bp

How would you know that your DNA band is the correct size?

To be able to estimate the size of DNA bands visualized by gel electrophoresis, we will have to load a 'ladder', a collection of DNA fragments of known size that we can use to compare our samples to.

The ladder that we will use in our experiment is called 100bp DNA ladder and is the collection of bands represented in the previous page.

- ✓ If your PCR experiment worked, where would you expect your DNA amplicon to migrate? Circle the spot on the ladder's figure
- ✓ Let's recap. What will you need for your gel electrophoresis experiment?

-Sample:

-Gel:

-Other:

Overview

In order to prepare your agarose gel, you will have to :

- Assemble a mould to pour the gel in. The mould will include the 'comb' that will leave wells into the agarose gel, once the gel solidifies. These wells will be used to deposit (load) the samples into the gel
- Prepare TAE 1X
- Prepare a 30 mL agarose gel in TAE
- ✓ For our gel electrophoresis, we will need ~400 mL of TAE, some to dissolve the agarose into a gel, and some to cover the gel itself and conduct electricity How would you prepare 400 mL TAE from a 50X stock? (2 pts)

- ✓ Once the TAE is done, you will have to prepare a 30 mL, 2% agarose gel in TAE (this means that the solvent is TAE, not water). What would you mix? (2 pts)

In order to run your sample into the gel, you will have to :
- Mix your sample with loading buffer
- "Load" your sample into the gel
- Assemble the gel electrophoresis chamber
- Apply electricity
- Wait for the sample to travel through the gel
- Take a picture under the UV light
- ✓ To load your sample into the gel, you'll need to mix it with loading buffer. The loading buffer is a 6X. You want to load 10 μL of your sample plus loading buffer. What will you do?

To load a PCR sample into a gel, inevitably, you have to open your PCR tube and pipet some of your sample out. This is the biggest source of contamination in any molecular biology/genetics lab.
- ✓ How many copies of your DNA fragment should there be in your tube after 35 cycles of PCR if you started from 10^6 cells (3 pt)?

For this reason the pipet you use to load the gel will be set aside for decontamination.
- ✓ What would you expect to happen if you were to prepare a PCR sample with the pipette you used to load your gel (1 pt)?

Gel loading tips

- Pipet slowly, avoid air bubbles
- If you aspirate all the sample and a bit of air, simply push down on the piston to expel the air. Maintain the pressure on the piston to keep the sample in the tip
- To make your hand steady, lean your right elbow onto the bench (if you are right-handed) and stabilize the barrel of the pipet with your left hand
- Release the sample SLOWLY into the well. If it looks like the sample is 'stuck' into the pipet it's because you jammed the tip in the bottom of the well. Simply lift the tip a bit and try again
- IF YOU RELEASE THE SAMPLE TOO FAST IT WILL OVERFLOW OUT OF THE WELL

Gel preparation

Experiment 5: Agarose gel prep

Materials

- Gel mould
- combs
- Sticky paper tape
- TAE 50 X
- Agarose powder
- balance
- parafilm
- beakers
- Spatulas

Protocol

1. Prepare 400 mL of TAE 1X by mixing 8 mL TAE 50X and 392 mL distilled water in a graduated cylinder. Cover with parafilm and invert to mix. Set aside
2. Weigh 0.6 g of agarose on the provided balance on a piece of parafilm and gently pour the powder in the provided flask
3. Measure 30 ml of TAE 1X and add it to the flask to obtain a 2% agarose solution.
    - ✓ Is this solution homogenous or heterogenous? (2 pts)
4. Let the agarose powder sit in the TAE buffer for a few minutes. This gives the agarose the time to hydrate and prevents the formation of bubbles when you microwave it. In the meantime assemble your gel tray by taping the sides and situating the comb
5. Cover the beaker with plastic wrap and bring it to the microwave
6. To avoid crowding, I will microwave the solution. This will be done slowly, monitoring to prevent over boiling. I will stop as soon as the solution boils, take it out, carefully mix it <u>with hot gloves on</u>, put it back, and repeat till the solution is completely clear
    - ✓ Is this solution homogenous or heterogenous? (2 pts)

7. Cool the agarose solution till it's hot but not scalding and pour the agarose evenly into the prepared mould. Eliminate any bubbles with a clean tip

8. Wait for the gel to solidify
    - ✓ Did it change color, once solidified?

9. Wrap your gel with Saran Wrap label it with your group's name and store it in the fridge till the next time

Gel electrophoresis

So far, you extracted your own DNA, quantified it by spectrophotometry, learned to design specific DNA primers, amplified by PCR the gene we wanted to analyze, TAS2R38, and prepared your agarose gel.

Now we need to see if the PCR worked. In order to do so we will run your PCR samples through gel electrophoresis, using the gel you prepared last time.

Objective

The objective of this experiment is to find out if your PCR experiment from last week worked.

Experiment 5: DNA gel electrophoresis

Materials

- power pack
- electrophoresis chamber
- Agarose gel prepared last time
- PCR samples
- TAE 1X prepared last time
- DNA 6x loading buffer with fluorescent molecule
- 100 bp DNA ladder
- parafilm
- pipettes and tips

Protocol

1. Unwrap your gel. Gently place it into the tray and into the chamber
   - ✓ Where should the wells' side be placed into the electrophoresis chamber? Toward the + or - electrode (1 pt)?

2. Gently pour TAE buffer over the gel and fill the side reservoirs. You want the gel to be completely covered by 2-3 mm of buffer, but not more. If air bubbles form into the wells, gently displace them with a disposable tip or ask for my help

    ✓ Why, do you think, you don't you want too much buffer on top of the gel (1 pt)?

3. I will show you how to load a sample into a gel by loading the ladders

    ✓ Each gel should have a ladder, why (1 pt)?

4. I will prepare 2 μl-droplets of loading buffer on a piece of parafilm paper. When your turn to load comes take 10 μl from your PCR sample, pipet up and down to mix the sample with the droplet of loading buffer and release your 10 μl sample into the well.

    ✓ Write below the order in which you loaded your samples, remember to include the ladder and the controls

5. Once everyone has loaded their samples, close the electrophoresis chambers under my supervision and start the run at 120V for about 15 minutes

6. At the end of the run, we will take a picture with the gel documentation. I will upload the pictures to Canvas. Please print it and tape in the results section properly labeled. You will also need to add that picture in your project report.

    ✓ What's your gel number?
    ✓ What's the order of the samples?

Gel Data Analysis

Last time you loaded and ran your PCR sample through gel electrophoresis.

✓ What is the goal of this experiment? (1 pt)

✓ What were the possible outcomes and what would they mean? (2 pts)

✓ Did your PCR reaction work?
✓ Did any PCR reaction work?
✓ Was the size of your fragment what you expected?
✓ Was the negative control clean?
✓ What do you conclude from your gel electrophoresis experiment?

✓ Remember to fill out the results at the end of the manual!

Amplicon purification theory

Gel electrophoresis gives you an indication about your PCR: did it work?

Yet, it does not really advance your project. It's more of a mid-point check. To find out your genotype at TAS2R38, we need to find out the DNA sequence. This will be done using the rest of your PCR reaction, which will be read via DNA sequencing. In order to do that, we will need to purify the amplicon, separating it from primers, nucleotides, and other PCR reagents.

Vocabulary

- **Silica** - a a thin layer of silicon oxide consisting of microporous structures that, in the presence of positive salts bind or release DNA (in size between 100-4,000 bp) depending on pH

- **Adsorbtion (noun, to adsorb, verb)** - molecules sticking to a surface as a thin film

- **Elution (noun, to elute, verb)** - remove an adsorbed substance by washing with a solvent

Overview

A PCR amplicon can be used for many applications: cloning, restriction digestions, or sequencing (literally learning the DNA sequence base by base. You will learn how this works later in the semester). In order to purify the amplicon from other reagents (primers, residual nucleotides, genomic DNA etc...) for further use, we will take advantage of **silica**. The silica membrane inside the columns I will provide has a high affinity for DNA —100 bp to 4,000— in a high concentration of salts, but not in low salt concentration.

✓ Would single nucleotides bind the silica? Why? 1 pt)?

✓ What about primers? Why (2 pt)?

✓ What about genomic DNA? Why? (1 pt)

Below is an overview of the general steps needed to purify an amplicon.
- Binding: PCR is mixed with a high salt solution with a pH indicator and forced to go through the silica membrane. At this acidic pH, the amplicon will bind, but all other impurities will flow through, including primers, which are too short to bind effectively the silica
- Wash: excess salts are washed away (DNA is still bound to silica membrane)
- Elution: An alkaline, low salt solution is added to the DNA, detaching the DNA from the silica membrane. The sample, containing only elution buffer and DNA, is eluted and quantified

Many kits are commercially available to purify amplicons. As you will see, we will add solutions with proprietary names, but in short, we will be performing the three steps above.

Once you have a purified amplicon, you can use it for DNA sequencing, the experiment that allows you to obtain a "physical genetic map," the literal DNA sequence of the amplicon, nucleotide by nucleotide.

DNA sequencing theory

Background

What is going to happen to the samples when we send them out for DNA sequencing? How can we "read" a sequence of DNA? It was only in the late 70s that Frederick Sanger developed a method to efficiently sequence DNA[6]. The method contributed to the awarding of his second (!) Nobel prize.

The method we use today is a modification of his original, brilliant idea.
It is based on PCR, but with a slight modification:

- The DNA to be sequenced undergoes 4 different PCR reactions in parallel.

- Each PCR reaction contains the 4 dNTPs, as customary, plus a modified nucleotide (ddNTPs, dideoxynucelotides triphosphate) that terminates the PCR reaction prematurely, because it does not allow further nucleotides to be tacked on

- These reactions are not exponential, because only one primer is used (as opposed to a primer set of forward and reverse. This is why you need to perform PCR beforehand, so that you have a lot of the sequence you want to read)

You didn't understand, did you? Can you imagine coming up with the idea in the first place? Let's try to understand how sequencing works. Let's look, for example, at the PCR reaction containing modified C.

In this tube, DNA polymerase will synthesize DNA as usual, but every time it runs into a G into the template strand it might randomly incorporate in the nascent strand either a normal C and go on with the synthesis, or a modified C, truncating synthesis.

Therefore, in the C tube, at the end of the experiment, there will be an array of fragments of different lengths, each one corresponding to the position of a C.

For example, if the newly synthesized DNA fragment called for Cs in position 3, 12, 25, 30, 32 etcetera, in the tube there will be fragments 3, 12, 25, 30, 32 etcetera bases long. Since DNA is complementary and antiparallel this information can be used to learn the position of Gs on the original sequence.

The "C" tube is only one of four. The fragments from all four tubes (modified C, modified G, modified A and modified T) are ran on four parallel lanes on a high resolution polyacrylamide gel, able to discriminate between sequences one base apart.

Combining data from the 4 lanes, each separating a PCR reaction carried out with one modified base, will give you the sequence of the nascent strand. Let's see how this works in practice...

Virtual DNA sequencing

We have four sequencing tubes. Each tube has all four nucleotides, but the "C" tube will also contain "defective" C, the "G" tube will contain also "defective" G etc...

Let's see what happens when "we" are given a template to sequence.

Normally a template needs to be at least ~200 bp, the primer alone will need ~20 bp to pair, but for simplicity I added a lot of dots to signify a much longer sequence and only 10 bp.

The template is:

5'-..........................GTGTCACAGT..................................-3'

Let's pretend we are DNA polymerase enzymes:
- ✓ In which direction will we always synthesize?

Let's see, only for those 10 bp, what would happen in each tube, one at the time. Start synthesizing. Every time you have to add a base corresponding to the modified base in your group you have the option to incorporate either a normal base and continue polymerization, or a modified base, which will cause you to fall off. Write below the fragments that will be generated in each tube:

- ✓ Tube A:

- ✓ Tube T:

- ✓ Tube C:

- ✓ Tube G:

✓ Now that our four reactions are complete, let's load them on a high resolution polyacrylamide gel able to separate fragments one base apart. Add a scale 1 to 10 with a one base resolution to the gel below. After the run, will smaller fragments be at the top or at the bottom of the gel?

✓ Look at your fragments in each "tube" and mark them with their base length
✓ Let's "load" tube A on the gel. Draw the bands you would expect at the end of the run
✓ Repeat for the other tubes
✓ Look at your base number 1. Is this the first base in our original sequence?
✓ Is it the last?
✓ What is it?
✓ What would we have to do to get our original sequence 5' to 3'?

✓ Consider the small sequencing gel in the figure below, and provide below it the original sequence analyzed

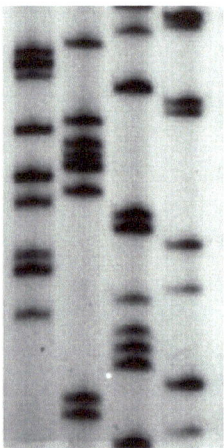

Need more practice?
- Make up a sequencing gel (14 bands) and an unrelated DNA sequence (10 bp)
- Send them to your assigned partner
- Based on their DNA sequence, draw what a sequencing gel would look like
- Based on their gel, come up with the original DNA sequence
- They will do the same for the gel and sequence you invented
- Discuss the results (and ask me if you are unsure!)

Amplicon purification

Experiment 6: PCR purification

Materials

- Buffer CP (column prep)
- Binding Solution
- Wash Solution
- Columns with DNA-binding resin
- 2-ml collecting tubes
- Water (elution)

Protocol

1. Insert a clean column (you can see the white silica at the bottom) in a collection tube. Label both with your lab number
2. Add to the column with 500 μl of CP buffer and centrifuge at 12,000 g for 30 seconds. Discard the flow-through
3. In a 1.5 ml tube, mix 200 μl (5 volumes) of Binding Solution and your remaining PCR (~ 40 μL). Add to the column
4. Centrifuge at 12,000 g for 1 minute. Discard the flow-through
   - ✓ Where is your DNA now? (1 pt)

5. To wash, add 0.5 ml Wash Buffer to the column and centrifuge for 30 seconds at 11,000 g
6. Discard the flow-through and place the column back in the same tube
7. Centrifuge the column once more for 2 minutes to remove any residual wash buffer

8. Place each column in a clean collecting tube. Label the top and side of the tube with your lab number followed by S (for sequencing)
9. To elute the DNA, add 50 μl of water to the center of the resin membrane
10. Let the column stand for 1 min, and then centrifuge as before
    ✓ Where is your DNA now? (1 pt)

11. Return the column to me
12. Quantify your eluate at the Nanodrop and record concentration and purity here and in the result section.

DNA sequencing prep

Objective

The objective of this part of the lab is to prepare your TAS2R38 PCR amplicon for sequencing to figure out which allele you carry (taster or non-taster?)

Materials

- Eluted amplicon (ideally 20-40 ng/μl)
- Sequencing Primer h-TAS2R38-DNA-R (P seq 2 μM)

Protocol

1. Ideally your concentration for sequencing should be 20-40 ng/μl. If your sample is more concentrated, dilute it to 25 ng/μl. If it is less concentrated or you got a negative reading we will try to sequence it anyway
   ✓ Did you have to dilute your sample? If yes, write your calculations here

2. Obtain a barcode tube and record your barcode below
   ✓ My barcode is:

3. Add to your barcode tube 8 μl of purified PCR product in the right concentration range and 4 μl of primer. Cap and put your sample in the sequencing bag

DO NOT FORGET TO PICK UP YOUR PTC BAGGIE!
This is the last time we'll see each other face to face!

DNA sequencing analysis and data discussion

Today, we will finally analyze your sequencing data to discover which alleles you carry at the TAS2R38 locus.

Vocabulary

- **Haplotype** - the combination of DNA sequences at adjacent loci on a chromosome, for example a sequence of alleles (A, b, c,d, E) or a sequence of SNPs

Background

TAS2R38 genetics

You already learned that the gene TAS2R38 encodes a membrane receptor associated with the ability to taste certain bitter substances.

There are three single nucleotide polymorphisms (SNPs) in the gene sequence that are known to result in three amino acids substitutions into the protein, affecting the phenotype.

These polymorphisms are:
-A49P
-V262A
-I296V

✓ Can you guess what the acronyms above mean (2 pts)?

The two common haplotypes are:

- AVI, non tasters

- PAV, tasters
   - ✓ Where do the acronyms AVI and PAV come from (1 pt)?

Heterozygous individuals (PAV/AVI) can taste the bitterness of glucosinolates, but to a lesser extent than homozygous individuals (PAV/PAV).

   - ✓ What type of dominance is this (1 pt)?

Following my instructions, determine which alleles you carry at the TAS2R38 locus.

   - ✓ What is your genotype? Write your answer below and in the results section

Experiment 7: Are you really a taster?

Now that we found out your genotype let's see if you really are a taster or a non-taster.

The paper that I am passing around has been spotted with PTC (phenylthiocarbamide), a glucosinolate. Rest assured that, in these amounts, PTC is harmless. Only, if you are a taster the paper will seem incredibly bitter.

   - ✓ Can you taste PTC (no, slightly bitter, yuck)? Add your answer in the results section

   - ✓ Is this consistent with your genotype? If not, let' discuss why together

   - ✓ Given the results of the whole lab, calculate the OBSERVED phenotypic frequency in the table below and add them to the results section

	Number of individuals	Phenotypic frequency
Tasters		
Non tasters		

✓ Given the sequencing results of the whole lab, calculate the OBSERVED genotypic frequency in the table below and add them to the results section

	Number of individuals	Genotypic frequency
TT		
Tt		
tt		

✓ Given the sequencing results of the whole lab, calculate the OBSERVED allelic frequency in the table below and add them to the results section

Calculated from genotype (seq)	Number of alleles	OBSERVED allelic frequency
T		p=
t		q=

We finally found out the allelic frequencies for the TAS2R38 gene for our population, but that was quite a bit of work! In some cases it's impossible to genotype populations (not everyone is willing to give their DNA, and can you imagine genotyping lions, for example?!?)

Yet, as you will learn at the end of the semester in lecture, allelic frequencies in a specific population are very useful for a number of predictions.

Once again, a brilliant solution came from thinking out of the box...

The Hardy Weinberg law: estimating allelic frequency based on the observed phenotype

✓ Imagine you hide in the woods and record a phenotype (10 white lions vs 45 brown lions). If brown is dominant over white, what would be in the way of finding the allelic frequencies? (2 pt)

In 1908 Hardy and Weinberg independently found an easy way around it.

Let's consider our population and call 'p' the frequency of the dominant allele B in the whole population (brown), and 'q' the frequency of the recessive allele b (white).

✓ Build a Punnet square for the entire population, following my explanation

✓ What is the genotypic frequency of BB, given the allelic frequencies p and q? (1 pt)
✓ What is the genotypic frequency of Bb, given the allelic frequencies p and q?
✓ What is the genotypic frequency of bb, given the allelic frequencies p and q?
✓ What is the phenotypic frequency for brown lions, given the allelic frequencies p and q? (1 pt)
✓ What is the phenotypic frequency for white lions, given the allelic frequencies p and q?
✓ Given what you wrote above, if you counted 10 white lions and 45 brown lions, what would be p and q in this population (5 pts)?

Let's try to apply HW to our taster/non taster population! <u>As you will learn later on, HW cannot be applied to humans (no random mating and several evolutionary forces are at work). Regardless, our sample would be too small.</u> Also, it is unclear if tasting glucosynolates is complete or incomplete dominance, which makes it a bit tricky, but let's try anyway for the sake of learning!

✓ Let's calculate allelic frequencies starting from phenotypic frequencies we found by tasting PTC paper:

OBSERVED allelic frequencies (calculated through sequencing)	OBSERVED Phenotype counts	OBSERVED Phenotypic frequencies	PREDICTED allelic frequency calculated through HW
	Tasters =		p =
	Non-tasters =		q =

- ✓ Are the OBSERVED allelic frequencies calculated through sequencing comparable to the PREDICTED allelic frequencies obtain through HW?
- ✓ <u>Copy your data in the results section of the manual and write your conclusions</u>

You are now ready to write your report 1: "Genetic analysis of the TAS2R38 locus; Are you a taster?"

Follow the instructions on the syllabus to describe the whole project including all of its experiments. Use the Project Overview and Results section to guide yourself.

Project 2 overview

Expression analysis: determining mRNA copy number

Objective

The objective of this virtual project is to understand RNA-based techniques (RNA extraction and retro-transcription), serial dilutions, standard curves, and real-rime RT-PCR. Using simulated data you will see how to go from spitting in a tube to determining how many copies of a specific mRNA (GAPDH) are in buccal cells.

Below is an outline by week (bullet points) with the experiments that will comprise this project.

This project will span several experiments over several weeks, as outlined below.

- RNA extraction and quantification as compared to DNA extraction
- Retro-transcription of RNA into double stranded cDNA: breaking the dogma
- Bioinformatics: design of expression analysis primers
- Real-time RT-PCR theory
- Serial dilutions
- Real-time RT-PCR data analysis

RNA extraction theory

So far we learned about DNA and DNA analysis.

Nonetheless, all the cells from the same organism share the very same DNA. It is the RNA that determines the phenotype—what a cell becomes and can achieve. For this reason, RNA extraction and analysis is of the prime importance in molecular genetics; which genes are expressed? How much?

Objective

The objective of this lab is to understand similarities and differences between RNA and DNA, DNA extraction and RNA extraction.

Vocabulary

- **Trizol** - a proprietary mix of phenol at the ideal pH for RNA extraction (from Life Technology)
- **Labile** - something that is not stable, changes easily or is easily degraded (like RNA!)

Experiment 8: RNA extraction from human cells, part 1

Materials

- Trizol (T tube in your rack)
- Chloroform (C tube in your rack)
- Glycogen (0.5 ml G tube)
- Isopropyl alcohol (I tube in your rack)
- 75% Ethanol (E tube, stored at -20 °C till needed)
- Molecular biology grade H_2O (H tube)
- 55°C water bath

  When working with RNA always use gloves and filter tips!

Background

RNA is a nucleic acid, like DNA.

While DNA lasts for the lifespan of the cell and is quite stable, RNAs are made and degraded continuously, their lifespan is typically very brief, and they are quite labile (they degrade easily). For this reason RNA should be handled with particular care. To avoid contamination with substances that might promote RNA degradation, no reagents to be used with RNA (including tips and pipets) are to be touched with bare hands.

RNA is transcribed using a template of DNA. Only a small portion of the RNA, comprised by mRNAs, is translated into proteins. Indeed, the bulk of the RNA of a cell is comprised by rRNAs and tRNAs.

Within the same organism, cells with different functions have a different set of RNAs, but identical DNAs. DNA can be used for genetic studies investigating the DNA sequence and the type of genes (alleles) present in an organism. RNA can be used to investigate what a specific cell is doing at a specific time, for example in response to a stimulus, or in comparison to another cell.

- ✓ Do you expect the DNA sequence of a cell from your liver to be identical or different from the DNA sequence of a cell from your eye (1 pt)?

- ✓ Do you expect the RNA pool of a cell from your liver to be identical or different from that of a cell from your eye (1 pt)?

RNA extraction is very similar to DNA extraction.

- ✓ What were the steps required for DNA extraction? (6 pts)

1.
2.
3.
4.
5.
6.

In order to extract the RNA you would have to:

1. **Collect** cells (for example by spitting in a tube, or collecting blood)
2. Break the cells' membranes to free the RNA (**lysis**)

    Cell lysis for RNA extraction is typically done using an affordable, proprietary mix called Trizol. Trizol breaks the cells and already contains phenol.

3. Separate RNA from DNA and proteins (**extraction**) by adding chloroform (it will form the aqueous phase).

We already discussed how to separate nucleic acids from proteins using phenol, when we learned how to extract DNA. Now we have to manage to separate DNA from RNA. The original method for RNA extraction was based on size differences between the two nucleic acids, and took about three days of centrifugations at very high speeds (ultracentrifugation) to carry out. Then in the 80s a mistake was made. Nicoletta Sacchi forgot to bring the pH of the phenol solution to 8.0, as was customary for DNA extraction, and ended up with a lot of RNA instead.

As it turns out, if the pH of the phenol solution is acidic, the DNA will end up in the organic phase, while the RNA will stay in the aqueous phase[7].

4. **Precipitate** the RNA (same principle as DNA precipitation)
5. **Wash** the RNA with ethanol
6. **Pellet** RNA and resuspend in molecular biology grade water
7. **Quantify** your sample at the fluorometer (like for DNA. For example you could use a Nanodrop or another spectrophotometer)

✓ What 260/280 ratio do you expect for pure DNA (1 pt)?

✓ What 260/230 ratio do you expect for pure DNA (1 pt)?

For pure RNA an ideal 260/280 ratio is a bit higher, typically **1.8-2.**

Retro-transcription

In order to later quantify a specific RNA species (one transcript) we will have to convert RNA into a more stable, double stranded form (cDNA or complementary DNA) that we can use for our "quantitative analysis".

Objective

The objective of this part of the lab is to learn how to convert RNA into cDNA, which can be used for further analysis (for example PCR and sequencing, which you are already familiar with, or like in our simulation, real-time RT-PCR analysis).

Vocabulary

- **cDNA**- complementary DNA. cDNA is a double stranded, DNA copy of RNA (it will represent only region of DNA that are transcribed and with no introns)
- **Hybridize**- to form a hybrid between, for example, a DNA probe and a complementary strand of RNA
- **Retro-transcriptase**- a viral enzyme able to synthesize cDNA from a template of RNA
- **Qualitative technique**- a technique that reveals the presence of a molecule
- **Quantitative technique**- a technique that reveals the amount in which a molecule is present
- **House-keeping gene**- a gene that is constitutive (always expressed)

Background

We want to investigate the expression of glyceraldehyde 3-phosphate dehydrogenase (GAPDH) mRNA in DYC students' buccal cells, but RNA is single stranded and very labile

- ✓ Would PCR work directly on RNA? Why (2 pts)?

Historically RNA was analyzed by **Northern Blot**, separating tiny amounts of RNA fragments into a gel and making the band of interest visible by hybridizing with a complementary radioactive probe.

- ✓ What could have been some problems associated with this technique (1 pt)?

- ✓ Where does the name "Northern Blot" come from?

- ✓ What is a Southern Blot?

- ✓ A Western Blot?

- ✓ An Eastern Blot?

The isolation of a viral enzyme, **retro-transcriptase** (RT), made the study of RNA much easier.
- ✓ What type of virus would use RT (1 pt)?

<u>RT synthesizes DNA starting from a template of RNA</u>, pretty much the opposite of RNA polymerase.

<u>Keep in mind that RT makes only one copy of cDNA per each RNA template as the RNA gets degraded in the process.</u>

- ✓ Is retotranscription an exponential process like PCR? Why? (2 pts)

- ✓ What's the advantage of using RT to convert RNA into its double stranded DNA correspondent, **cDNA** (complementary DNA) or better, what can we do with cDNA that we could not do on RNA (1 pt)?

- ✓ Which enzymes have you encountered in the lab so far, and what are they used for in the lab (3 pts)?
  -
  -
  -

Like DNA polymerase, RT needs primers and bivalent cations to work.

If you were interested in analyzing only the expression of a specific gene, you could design mRNA-specific primers.

More frequently scientists convert all RNA in a sample into cDNA.

- ✓ Can you imagine a type of primer that would allow you to retro-transcribe all the mRNAs (5 pts)?

- ✓ How about primers to retro-transcribe all RNAs, including mRNA (5 pts)?
- ✓ To summarize, what are the three possible types of primers you can use in retro-transcription? (3 pts)
    - ◉
    - ◉
    - ◉

- ✓ Do you think that cDNA would be more, less, or about the same amount of the original RNA present in a sample (2 pts)?

Either type of primer can be used in RT reactions.

cDNA stands for complementary DNA, and is a library of all RNAs expressed in a cell at the moment of the RNA extraction. Keep in mind that RT does not AMPLIFY! Depending on the enzyme used, either a complementary strand is added or a the original RNA is also replaced, but in all cases one copy of RNA leads to one cDNA.

- ✓ Would cDNA generated from a population of buccal cells be analogous to that generated from a population of liver cells from the same individual (1 pt)?

- ✓ Would DNA generated from a population of buccal cells be analogous to that generated from a population of liver cells from the same individual (1 pt)?

- ✓ Why is that different?

Once you obtain cDNA from a population of cells you can analyze it by PCR, similarly to what you did for DNA, only your primers will need to be designed a bit differently. Also, the PCR technique you already learned is qualitative, if you obtain a band then the PCR worked and the gene was amplified. Next week we will learn how to make PCR quantitative, meaning we will learn exactly how many copies of RNA we started from.

Overview: Retro-transcription of RNA into cDNA (RT)

Now that you became a little more familiar with molecular genetics, can you guess the ingredients you will need in your RT reaction? Keep in mind that RT, much like DNA polymerase, also needs bivalent cations. To set up an RT reaction you will need:

- ✓
- ✓
- ✓
- ✓
- ✓
- ✓

Ideally RT reactions start from 100 ng-5 µg of total RNA.

Next let's see what steps would comprise a retrotranscription reaction. The steps below are based on "The Firsto-strand synthesis kit" by Promega. Different companies will have slightly different protocols.

RNA is full of secondary structures. In the figure aside (from the public domain), you can see the secondary structure of an example RNA.

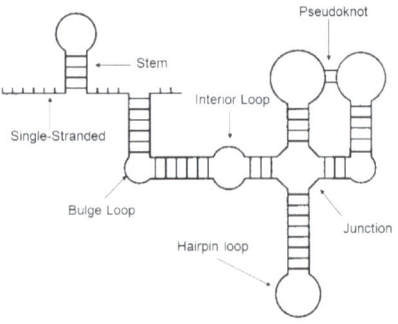

RNA Secondary Structure

✓ Do you think the primers will be able to pair to their complementary RNA, if the RNA is full of secondary structures?

✓ How could we break the secondary structure (1 pt)?

✓ What would we need to do, then, to allow the primers to anneal?

- Indeed, you would first mix the RNA with the primers, denature at 70 °C for 5', and put the sample on ice.
    - ✓ What will happen when you put the samples on ice?

✓ At what temperature did we denature DNA when we did PCR? (1 pt)

✓ Why do you think the temperature is different for retrotranscription? (1 pt)

- Then you will add all the remaining reagents and incubate at 25 °C for 10', 42 °C for 1 hour, and 70 °C for 15'

    ✓ Let's summarize, add duration, and write what the different temperatures are for

    - 70 °C
    - 0 °C (ice)
    - 25°C
    - 42°C
    - 70°C

✓ A typical RT reaction is carried out in 20 μL. With this information complete the table below.

Order and ✔	Reagent	1 reaction
	RNA	10.8 μl
	Random hexamer	1 μl
	MgCl$_2$ (25 mM)	1.2 μl
	RT Buffer (5x)	
	dNTPs (10 mM)	1 μl
	Rnasin (RNase inhibitor)	1 μL
	RT	1 μl
	Tot volume	

✓ Given the stock concentration (in the table) of the MgCl$_2$ and dNTPs, calculate the working concentration of both solutions in an RT reaction

✓ In which order should you add the reagents? Write it in the table below (left column)

✓ Assume that you extracted your RNA and obtained a concentration of 3.4×10^4 ng/mL. If you added 10.8 μL of RNA, how many ng of RNA did you add to the RT? (2 pts) Write the answer in your virtual results for project 2.

✓ Given your answer above. If the total RT volume is 20 μL, what's the new RNA concentration within the RT reaction? (2 pts). Write the answer in your virtual results for project 2.

✓ What are differences and similarities between RT and PCR reactions?

Bioinformatics: expression-primer design and RNA review

So far you virtually extracted RNA from your buccal cells, quantified it, and converted it into cDNA. This week we will design a primer set to analyze the expression of a specific mRNA: glyceraldehyde 3-phosphate dehydrogenase (GAPDH).

Objective

The objective of this part of the lab is learn to design primers specific for the analysis of cDNA.

Vocabulary

- **Isoform or variants**- alternatively spliced RNAs originating from the same gene
- **Exons**- the parts of an eukaryotic pre-mRNA that are kept after splicing
- **Introns**- the parts of an eukaryotic pre-mRNA that are removed during splicing

Overview: designing expression analysis primers for GAPDH

Genes are "expressed" when RNA is transcribed using DNA as a template. Otherwise, genes are said to be "silent" or "not expressed". Primers designed for expression analysis (to assess presence and/or calculate copy number of one specific RNA) are different than those designed for DNA PCR.

✓ Do you remember what's the function of the enzyme glyceraldehyde 3-phosphate dehydrogenase (1 pt)?

- ✓ Do you expect it will be expressed in your buccal cells (1 pt)?

- ✓ Do you expect every RNA to be expressed in every cell (1 pt)?

- ✓ Do you expect every gene to be present in every cell (1 pt)?

- ✓ Would it be a problem if your primer set amplified not only cDNA, but also genomic DNA? Justify your answer (1 pt)

- ✓ To think of a way to avoid that let's list the main differences between a gene and its corresponding cDNA:

- ✓ Can you think of a way to design primers that only amplified cDNA, but not genomic DNA (1 pt)?

In order to design specific, effective primers for expression analysis we will have to:
- Look into databases to retrieve the mRNA sequence (not the genomic sequence!) of the transcript we are interested in
- Select the region where we want to design primers, such as to include two different exons, possibly separated by a very big intron
- Design the primers, so that they are specific (amplify only the transcript of interest) and effective (primers that are 100% efficient will double the number of amplicon at each round of PCR)

    We are going to design a primer set for GAPDH mRNA. We want our 'amplicon' (the region copied by PCR) to be about 100-300 bp, and to include a big intron (or more than one) in the corresponding DNA sequence.

<u>The length of expression analysis amplicons depends on the application.</u>

For traditional retro-transcription and PCR followed by gel electrophoresis, 200-600 bp is ideal.

For quantitative (real time) PCR, which is the quantitative technique we will carry out next week, an ideal amplicon is 100-300 bp.

- Double check the specificity of the primers designed

Experiment 11: Bioinformatics, expression analysis primer design

Materials

- A computer

Take copious amounts of notes as the expression analysis primer design that we are going to carry out today is almost identical to what you will have to do for your second primer design assignment

Protocol

1. Google gene cards and click on www.genecards.org
2. On gene cards, search for GAPDH and click on the correct gene card
3. Click on the 'transcripts' section, and under the very first mRNA/cDNA section click on the last RefSeq RNA sequence (NM_002046, variant 1)
4. Click on "Go to" and select "features"

   ✓ How many exons are in this transcript (1 pt)?

   ✓ At what position does the coding sequence (CDS) start (1 pt)?

   ✓ If you click on CDS, NCBI will show it to you. Try, then go back by clicking feature on the bottom left of the screen

   ✓ Can you infer the size of the exons? For example, how long is the first exon (1 pt)?

   ✓ Are all the exons coding (3 pt)?

   ✓ Can you infer the size of the introns (1 pt)?

5. Right click on "Pick primers" in the top right corner and open it in a new window
6. This time we don't need the primers to be in a specific position, as long as one or more big introns separate them. Ignore the PCR template box
7. Under the Primer Parameter box input the desired size of your PCR product (amplicon). Today we need it to be between 100-300 bp (your assignment asks for a different size. Check on the syllabus!)
8. In the same box type "10" as your desired number of primers (primer pairs, really) to return (otherwise it gives you back 10 possible pairs and you might get confused. They would all be suitable)
9. Under the Exon/Intron selection box select under "Exon Junction Span" that Primer must span an exon/exon junction. This is ideal! A sequence spanning across an exon joint will only exist on a mature mRNA. (If on your home assignment the software returns no primers, uncheck this option as sometimes it can be too limiting. If this solves the issue write in the assignment WHY you unchecked it, or you will lose points!)
10. Checkmark the "Intron Inclusion" box (If after unchecking the exon joint inclusion you still get no primers, uncheck this too. If you were to proceed with a primer pair designed this way you would have to include a no RT sample as a control in your PCR. If you are curious ask me about this. If not, forget it)
11. Checkmark "show results in a new window" and click "Pick primers"

  Should you get no results in your home assignment, read the parentheses in points 9 and 10

  ✓ Did you find primers? If yes, name the primers h-GAPDH-RNA-RT-F and R. Write the sequences (5'to 3'!) in your results' section!
  ✓ What's the amplicon (product) size (1 pt)?
  ✓ What's the total intron size included (1 pt)?

12. Let's double check for the specificity of this primer set. Google MFEPrimer 3.0 in a new window and open it. Scroll down and under "online servers" click MFEPrimer 3.01
13. Copy and paste your two primers
14. Select Human GRCh37 RNA as your background database
15. Select a minimum Tm of 50 °C
16. Hit Submit

- ✓ Does your primer set amplify anything?

- ✓ Does it recognize the correct transcript? How can you tell (1 pt)?

- ✓ Is it specific (1 pt)? Why?

- ✓ Is the size of the amplicon what you expected (1 pt)?
- ✓ Write your findings in the results section

Real time PCR theory

Objective

The objective of this part of the lab is to understand how real-time RT-PCR works.

Vocabulary

- **Ct** - The threshold cycle at which quantitative (real time) PCR analysis is carried out
- **Primer efficiency** - how efficient is a primer set at amplifying its target transcript. An efficiency of 90% means that the primer set amplified 90 copies of transcript for every 100 copies that were present in the sample
- **Primer dimer** - primers annealing to each other during a PCR experiment and generating a short, aspecific amplicon
- **Melting curve** - an increasing temperature gradient performed at the end of a real-time RT-PCR experiment to monitor amplicon dissociation

Background

To find out if GAPDH is present in your sample we could just run a PCR, very much like we did for our genomic analysis, and run the result on an agarose gel.

✓ What would be the possible outcomes of such an experiment (2 pts)?

✓ What controls would you need?

But how do we find <u>how much</u> of a transcript is present? This type of analysis is very useful to <u>compare different cell populations</u>: for example, liver versus stomach, cancer versus normal tissue, non-treated cancer versus cancer treated with a drug, etcetera.

Remember that <u>the phenotype of any cell is determined by which genes are expressed and in which amounts.</u>

So, how do we quantify transcript levels?

Higuchi and his collaborators found a brilliant way to quantify transcripts in the 90s[8], which eventually became perfected in the technique we now know as real-time (or quantitative) RT-PCR.

The principle of quantitative (real time) PCR in its many variations is to <u>add to the reaction a substance emitting a signal proportional to the amount of double stranded nucleic acid present in the sample, while monitoring the reaction continuously (in real time)</u>.

Basically, as DNA polymerase (Taq) synthesizes double stranded DNA (dsDNA) in an exponential fashion, the signal will also grow exponentially.

✓ What controls, do you think, you will need to add in your real-time experiment (2 pts)?

During the exponential amplification phase, the signal will be proportional to the initial amount of cDNA. Of course, you will need a standard curve to determine the correlation between fluorescence and copies of a specific cDNA.

✓ What is the formula that tells you how many copies of nucleic acid you have at the end of PCR, if you start from a copy number "a" and go through "n" cycles of PCR (2 pts)?

✓ Fill in the table below with how many copies of cDNA you would have after 2, 3, 4, 10, 20, 30, and 40 cycles of PCR, if you were starting your experiment with 10^3 copies of the specific cDNA you are trying to amplify.

Number of cycles	Formula	Copy number	Log copy number
0		1000	
2			
3			
4			
10			
20			
30			
40			

✓ Could you easily fit the hypothetical results above in a linear graph? Why (1 pt)?

✓ Can you think of a way around that (2 pts)?

✓ Write in it the last column of the table above on the right the logarithm (base 10) of the copy number

✓ Draw below a graph representing Cycle number on x axis and the Log of copy number on the y axis

✓ The logarithm of the signal, of course, is not exponential anymore. Instead it's... (1 pt)

✓ How much did the copy number (not the Log) increase between cycle 2 and 3 (1 pt)?

✓ And between 3 and 4?

✓ Is this making sense with what you know about PCR? Explain (1 pt)

Real-time PCRs are ran on instruments that can monitor the intensity of the signal continuously, hence the name "real-time".

Imagine now that you are running at the same time, in the same conditions, six samples that you want to compare. The figure below shows what a typical real-time experiment looks like:

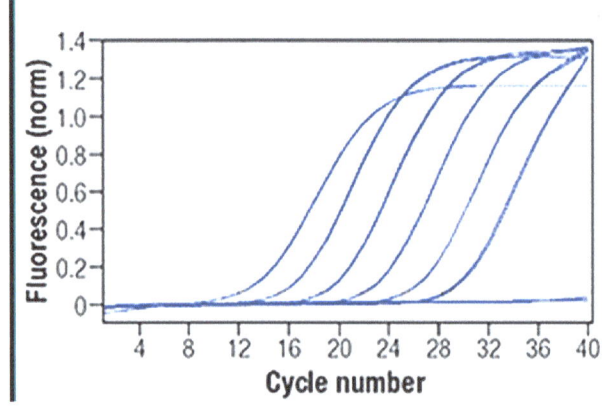

A typical real-time PCR result graph

✓ What does each curve represent (1 pt)?

✓ Name the samples in the figure above "A" to "F", left to right
✓ What is on the x axis?

✓ What is on the y axis?

✓ Why isn't fluorescence increasing before cycle 8/12 (3 pts)?

✓ Is the copy number increasing before cycle 8/12 (2 pts)?

✓ Why does the signal stop increasing between cycle 20 and 40 (3 pts)?

✓ Why are curves shifted as compared to each other (5 pts)?

CHALLENGE!

✓ Which curve on the graph represents a sample that has a higher initial copy number of cDNA (5 pts)?

✓ If you were to compare the samples, which portion of the curve would you use (2 pts)?

Indeed, that is how the analysis is done. A horizontal line (**threshold**) is drawn across the samples, as low as possible while still being in the exponential amplification phase (linear in a Log graph, like above) for all the samples.

The intersection between the threshold and each sample is called the threshold cycle (Ct) and is the value used for real-time analysis.

✓ Draw the threshold in the graph above and fill the table below

Sample	Ct
A	
B	
C	
D	
E	
F	

✓ Which sample had the highest initial amount of DNA or cDNA (2 pts)?

✓ You ran two cDNA samples in real-time RT-PCR with primers for GAPDH. You draw the threshold and see that their Ct value is only 1 cycle apart. Does that mean that the content of GAPDH RNA was very similar in the two samples? How different was it (3 pts)?

✓ If you were to run on an agarose gel samples A through F at the end of your real-time experiment, which sample would have a fatter band (2 pts)?

There are two main types of reagents used for real-time detection. Both emit fluorescence proportionally to dsDNA, but the first type consists of probes that binds specifically the target gene. These are very specific but more expensive.

The second type are substances like **SYBR green**, the fluorescent molecule we will use in our experiment. Its signal is proportional to ANY double stranded DNA.

✓ What could be the drawback of using SYBR green (1 pt)?

Initially, to make sure that SYBR green real-time signals were specific, samples were ran onto agarose gels at the end of the real time experiment. This was obviously very labor intensive.

✓ What would you expect to see on an agarose gel for a real-time sample that was specific (3 pts)?

✓ And what would you expect to see for a primer set that was not specific (1 pt)?

Once again, a brilliant idea solved the issue.

✓ SYBR green signal is proportional to the amount of double stranded DNA and is monitored continuously inside the well. What would you expect would happen to the fluorescence signal if, at the end of the real time experiment, you gradually increased the temperature up to 95°C (3 pts)?

✓ If your amplification were specific, what would happen to fluorescence once you reached the Tm (2 pts)?

✓ Draw a graph with the temperature on the x axis and fluorescence on the y axis. Then draw what a 45 cycle real-time PCR sample (A) that had a specific amplicon with a melting temperature (Tm) of 83°C would look like as temperature increases

✓ Add on the graph above a sample (B) that was positive for the same amplicon, but had also primers dimers (primers annealing to each other forming a short, aspecific amplicon). The melting temperature for the primer dimer will be 35°C in this example

What you just drew above is a **melting curve**. The melting curve you will see looks more like a bump because it plots the change in fluorescence (one big drop equals one big bump, se figure beside). If there is only one amplicon, the curve has only one bump at a temperature specific for that amplicon. If there are more bumps to the curve, there are aspecific amplicons, and the sample should not be considered for data analysis.

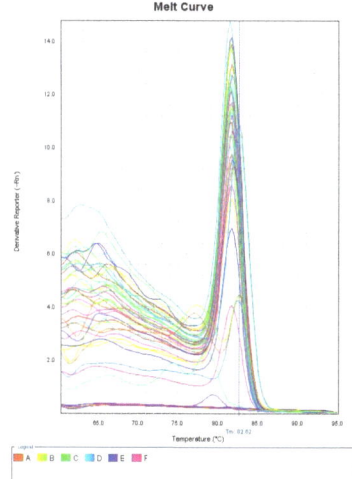

While comparing real-time curves ran within the same experiment might give you an idea of their relative amount of initial cDNA (2.3 times, 5.6 times etc…) they give you no information on the <u>absolute amount of cDNA</u>.

✓ Why?

Also, you would not be able to compare the expression of transcripts analyzed with different primer sets. Let's see why …

Assume you designed two primer sets; one amplifies cDNA for GAPDH1 and the other for GAPDH2 (two isoforms of GAPDH that undergo differential splicing).

You want to compare their expression in your buccal cells. You run a real-time RT-PCR and find a Ct value of 15.3 for GAPDH1 and 16.3 of GAPDH2.

You conclude that the transcript for GAPDH1 is double the transcript for GAPDH2. Unfortunately, you did not know that the primers for GAPDH2 don't work very well and only amplified one copy every two of the transcript (their efficiency was 50%).

- ✓ If you had that piece of information, what would you have concluded from your experiment (2 pts)?

For these reasons, the most accurate way to carry out a real-time experiment is to build a **standard curve**; because you know how many copies of cDNA are in each of your standards, you know how many copies each primer set should detect, hence you can calculate their efficiency.

Real-time standard curves typically are built with 6-8 standard points, each ten times more concentrated than the next, like in the example below.

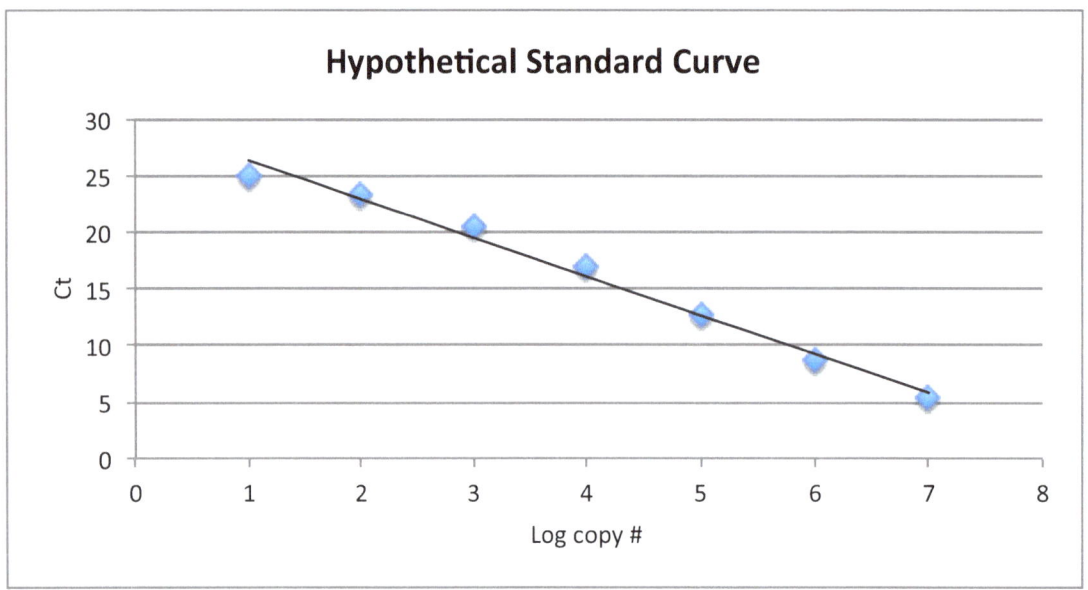

- ✓ Look at the first experimental point on the left. What is its Ct value? How many copies of RNA did that correspond to (2 pts)?

- ✓ For the standards, which of the two informations above is found during the real time experiment (2 pts)?

- ✓ Which one was already known?

- ✓ Given the graph above draw (guesstimate) corresponding real-time RT-PCR curves as the ones on page 101

The values of the standards will have to include the range of your samples, and should be determined accordingly. If your sample is outside of the range covered by your standards, you cannot assume the relationship between concentration and single is linear.

For example, the figure below shows the relationship between BSA absorbance and protein concentration, which you used (or will use) in your biochemistry lab

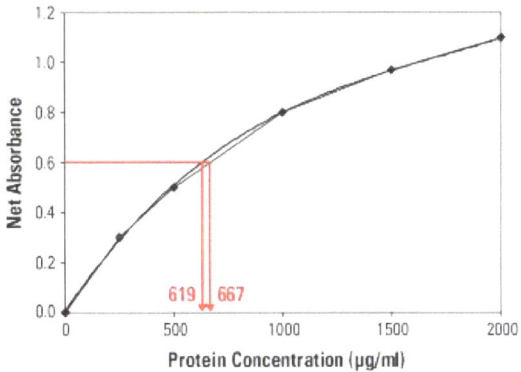

As you can see, the relationship is linear only for absorbance values below 0.6!

For real-time, typical standard values range from 10^3-10^{11} copies/reaction.

Next time we will discuss serial dilutions, which we will need in order to prepare our standards and run the real-time experiment.

Serial Dilutions

Objective

The objective of this part of the lab is for you to understand and perform correctly serial dilutions, at least on paper. We will do it practically the next time to prepare your real-time standards.

Background

The best practice to prepare standards covering a wide range of concentrations is—rather than prepare each standard from a stock—to perform **serial dilutions**. "Serial dilution" means dilution in a series; you will start by preparing the most concentrated sample from a stock and then use that standard to prepare the next dilution, like shown in the example below (copies stands for copies of the cDNA you want to quantify).

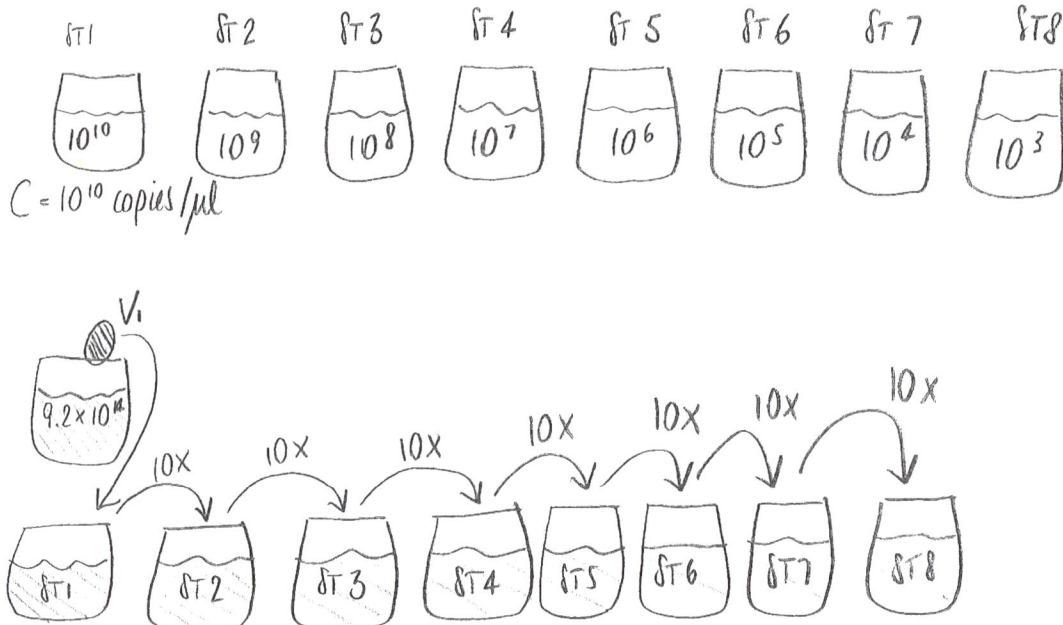

Serial dilutions. It is desired to prepared a series of samples with the concentrations specified in the top part of the figure. Rather than preparing each one of them from a stock, a serial dilution, like presented in the bottom part of the figure, saves time and is more precise.

The trick to serial dilutions is to keep the same dilution factor (C_1/C_2) from one sample to the next, for example 10x in the figure above.

Because each solutions is ten times less concentrated than the next, for each step of the dilution you can transfer 10 µl into 90 µl of water, mix, and repeat for the next step. This will save a lot of time and will make your dilutions more precise.

- ✓ GROUP DISCUSSION: For example, you need to prepare your real-time PCR standards. You want to pipet at least 5 µL of standard in each real-time reaction (to avoid imprecisions due to smaller volumes) and end up with 10^3, 10^4, 10^5, 10^6, and 10^7 copies of GAPDH in each reaction. At which concentrations do you need to prepare your standards?

- ✓ Which standard should you prepare first, the most or least concentrated? Which one is that (2 pts)?

- ✓ Any idea on how could you come up with an amplicon stock to prepare standards? (5 pts)

- ✓ Remind me, where does the stock come from (1 pt)?

- ✓ Assume the stock concentration is 4.88×10^7 copies/µl. Calculate how to prepare your first standard (3 pts). Assume a final volume of 100 µL.

✓ Fill in the table below:

Standard #	Standard concentration	µl of water	µl of sample
1	0 copies/µl	100	0
2	2×10^2 copies/µl	90	10 of st#3
3	2×10^3 copies/µl	90	10 of st#4
4	2×10^4 copies/µl	90	10 of st#5
5	2×10^5 copies/µl	90	10 of st#6
6	2×10^6 copies/µl	_____	_____ of stock

Serial dilutions: a survival guide

1. Prepare the most concentrated standard from the stock using $C_1V_1 = C_2V_2$
   - You typically will have C_1 and C_2, but not V_1 (which you have to calculate) or V_2. Remember, V_2 is how much of the standard you want to prepare, which will include how much you need in your experiment PLUS what you will use to prepare the next standard down the serial dilution line. Trick? Abound. Unless it's expensive, if you need 100 µl prepare 200 µl. If you need 1 mL prepare 2 mL. If in the next step you realize you don't have enough, that's okay! Go back to your calculations and double everything.

2. Calculate the dilution factor of the serial dilution
   - This is the ratio between the concentration of standard 1 (most concentrated) and standard 2 (next in line). It should be the same if you do 2/3, 3/4, 4/5 etc...

3. Calculate the fixed volumes of your serial dilution
   - If your dilution factor is 10 then in all of your successive standards 1/10 will come from the previous sample and 9/10 will be water. All the standards will be prepared the same way!

4. Double check! Are you left with enough volume of each standard?
   - You took out some standard to make the next one and now you don't have enough to run your actual experiment. Don't panic! That's why you wrote it on paper first. Just double or triple your fixed volumes, as needed.

TIPS!

- Always plan your dilutions on paper first!
- Always pipette volumes that are at least 5 µl
   ✓ Why would you want to avoid small volumes?

- Always label all of your tubes beforehand
- Plan for more! If you need 80 μl of a solution prepare at least 100 μl, 200μl if it's inexpensive
- Remember to pipet <u>into the solution,</u> not on the wall of the tube, don't have someone else hold the tube while you pipet, mix well but gently with the pipette, and collect all the solution at the bottom of the tube before proceeding to the next dilution
- <u>It is essential that you take all possible precautions to avoid sample contamination.</u> Realize that standards contain millions of copies of exactly what you are trying to measure by real-time PCR, which is extremely sensitive. <u>Avoid 'popping' open standard tubes. Pipet slowly and carefully. Avoid pipets touching the inside of standard tubes.</u>
    - ✓ What do you use to prepare real time standards?

Exercise: Practicing conversions and serial dilutions

- You run a standard PCR for GAPDH in 50 μl, then run a gel electrophoresis and find that your experiment worked (YAY!). You purify the amplicon (like we did when we prepared TAS2R38 for sequencing) and find that the GAPDH amplicon (300 bp) concentration in your sample is 14.3 ng/μl, which corresponds to 4.42×10^{10} copies/μL
    - ✓ How would you prepare your serial dilutions if you wanted standards from 10^3 to 10^{10} copies/μl? You will need 15 μL of each standard to run your real time PCR.

- ✓ What if you wanted to prepare standards from 10^1 to 10^7 copies/μl? (Keep in mind this is one of those instances you really do not want to pipet less than 5 μl. Can you imagine how bad it would be if your own standards were imprecise?)

- ✓ The next week you want to prepare BSA standards for protein quantification. The concentrations you want to cover are between (more or less) 100-1000 μg/ml. You want at least 4 standards (five is fine, too) and your stock is 2 mg/ml. What do you do? You need 50 μL per standard.

- ✓ How would I prepare standards at 5, 25, 125, 625 ng/mL from a stock at 1000 ng/mL? I need at least 50 μL of each standard

✓ How would I prepare standards at 3, 9, 27, 81 ng/mL from a stock at 0.5 ng/µL? I need at least 100 µL of each standard

✓ How would I prepare standards at 10, 20, 40, 80 mg/mL from a stock at 300 mg/mL? I need at least 60 µL of each standard

Real time PCR

Overview

✓ Why is this technique called real-time RT-PCR and not just real-time PCR (2 pts)?

Real-time RT-PCR is so sensitive that, to minimize pipetting errors and the likelihood of contamination, all the reagents come already assembled into a 2x master mix. The kit used for this protocol is the EXPRESS SYBR™ GreenER™ qPCR Supermix, with premixed ROX from Thermo Fischer Scientific. Make sure to adapt your protocol to your kit of choice.

✓ What reagents will be in a real-time master mix? (HINT: think about everything that goes in a PCR plus the one special ingredient in real time) (5 pts)
    ✓
    ✓
    ✓
    ✓
    ✓

✓ Which reagents will you have to add yourself, based on the experiment (2 pts)?
    ✓
    ✓

✓ Imagine you want to quantify GAPDH in your cDNA sample. To do that you would need standards (for example 10^3, 10^4, 10^5, 10^6, and 10^7), your sample, and which controls? (1 pt)

✓ If you want to run technical triplicates (which is the norm for real time) how many samples would you need in total? How many samples should your master mix include?

✓ As you can see below, real-time master mixes are very simple, to minimize error. Fill the table below for your virtual number of samples.

Reagent	1x	____x
2x premix	7.5 μl	
Primers (750 nM)	2.5 μl	
Total	10 μl	

✓ Below is what a real-time plate with 48 wells look like. Is it familiar? You learned to pipet using one! Let's make a plan as to where you would load each sample.

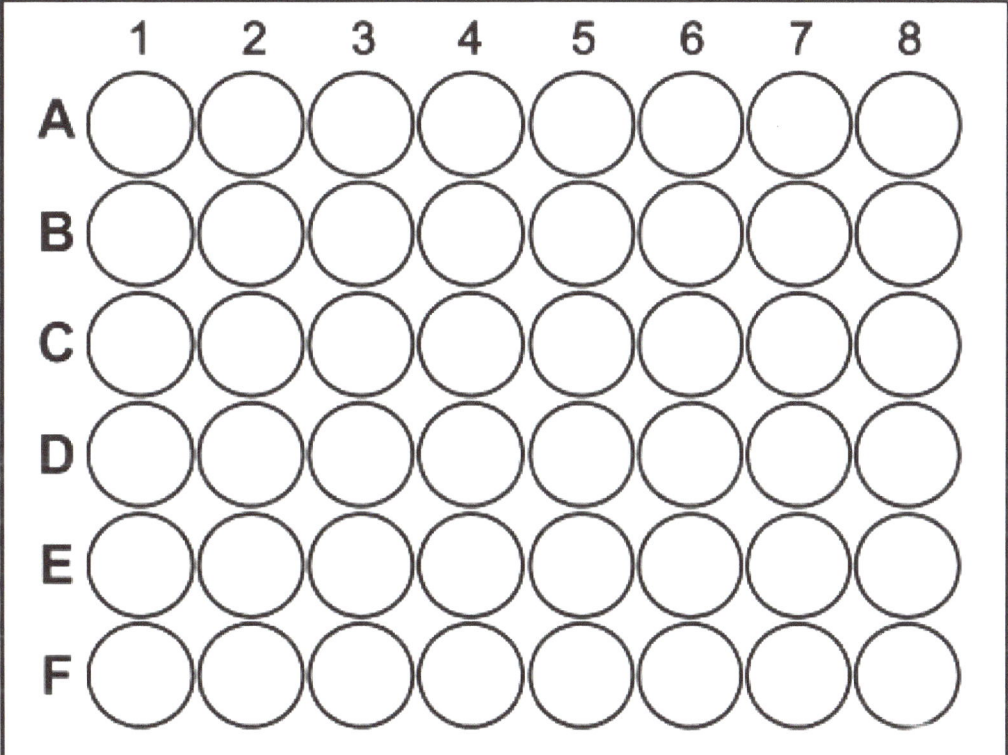

✓ What would you need to pipet in each well?
    1.
    2.

Real time plates are sealed with an adhesive to prevent well cross-contamination and then ran into a real-time machine.

Real-time data analysis

Overview

The real-time PCR machine displays data in a graphic format, like shown in the figure below.

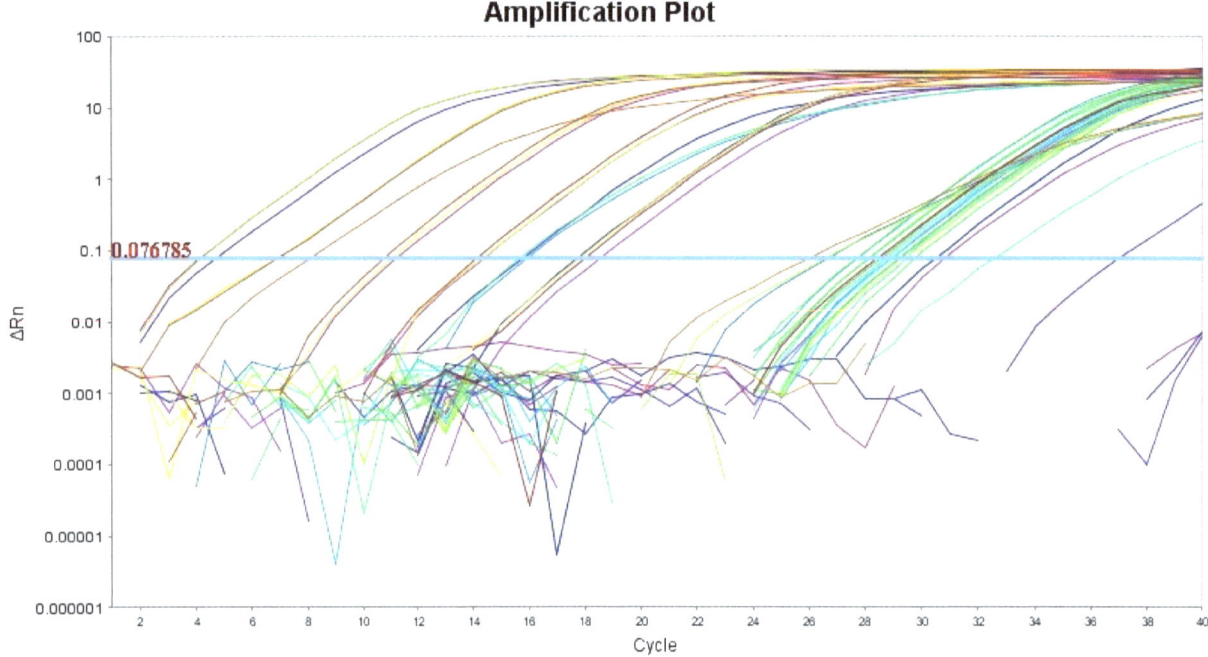

- ✓ What's on the x axis (1 pt)?

- ✓ What's on the y axis (2 pts)?

- ✓ Is the y axis logarithmic or linear (1 pt)?

- ✓ What's the horizontal line crossing the samples (3 pts)?

- ✓ Do you recall what is the definition of Ct? Write it below (2 pts)

✓ By looking at the graph, what is the Ct value of the second sample on the left (1 pt)?

✓ The last sample on the right (1 pt)?

✓ What are the squiggly lines below the threshold (2 pts)?

Once you set your threshold (the machine can choose it automatically, or you can pick it manually), the machine returns your results in an excel format, with all the Ct values associated to the samples on the plate. Those are the files that we are going to work with today.

To determine the copy number of GAPDH mRNA in your cDNA samples, we will need to calculate the efficiency of the GAPDH primer set we designed.

✓ If you your primer set had an efficiency of 100%, how many cycles apart would standards diluted 10 times be (5 pts)?

✓ Assume that your 10^8 copies/reaction standard has a Ct=8. Fill in the table below with Ct values for the other standards assuming an efficiency of 100%

Log (copy number)	Standard copy number	Ct
	1E+03	
	1E+04	
	1E+05	
	1E+06	
	1E+07	
	1E+08	8

✓ Add the logarithm (base 10) of the copy number of your standards

- ✓ Build a graph with the Ct (y axis) and the logarithm of the copy number of your standards (x axis)

- ✓ If you wrote an equation for the regression line describing your standard curve, the slope would depend on the increase of signal from one sample to the next in the dilution series. Therefore the slope would be (3 pts)…

The efficiency of your primer set will rarely be 100%, meaning that your slope will rarely be -3.32. To calculate your primer efficiency apply the formula:

$$\text{Efficiency} = 10^{(-1/\text{slope})} - 1$$

- ✓ Calculate the efficiency of your primer set if the slope of your curve is -3.5

Generally, only primer sets with an efficiency of 90% and above are selected for further experiments.

Protocol: real-time data analysis

1. Open the data file in Numbers or Excel
2. Locate your samples onto the spreadsheet and label them in the 'sample' column
    - ✓ What is the Ct value of your blanks? Write it below and in the results section
    - ✓ Did you expect a Ct value in your blanks?

3. The 'Tm' columns (Tm1, Tm2, Tm3…) report data from your melting curves
    - ✓ If your amplification were specific, how many Tms would you expect per well (1 pt)?
    - ✓ What can you conclude when there is a Tm2 listed (1 pt)?

- ✓ Did any of your wells show multiple Tms? If yes, write the well name and its corresponding sample below and in the results section

- ✓ Given that you amplified the same amplicon in all of your wells, did you expect the Tm1 to be consistent among wells? Was it?
- ✓ What is the definition of mean?

- ✓ What's the formula to calculate the mean in a set of values? (2 pts)

4. Let's calculate the average (or mean) Ct between your duplicates. Since you are analyzing technical replicates, you expect them to be identical. Any variation in Ct value is due to either technical error or to the machine's precision. Assuming no major technical error occurred, (you'll learn how to spot if that was the case) the average is closest to the true value., The more replicates, the most accurate your average.

   Click on the cell where you want your average to appear. Click the equal sign to open the formula dialogue box. Select average. Click on the values you want to average and hit return. Select the calculated average and drag the dot at the bottom of the cell down, to apply averages to the other duplicates

5. Hover on the Mean Ct you just calculated. Click on the little yellow (or purple or white) dot (or square)and drag it down for the other values you want to calculate the average for

   - ✓ Look at your duplicates, do they seem close? What are two statical values that estimates how spread are samples within a group?

   - ✓ What is the formula to calculate them? (2 pts each)

   Standard deviation (SD) tells you how far apart are your replicates. An ideal SD among replicates in a real-time PCR experiment is below 0.2. Ct values that are very high (very low copy number) will have higher SDs (the precision of the nmachine teeters at low and high Ct values). **Typically SDs below 0.5 can be attributed to the machine's precision. If your SD is above 0.5, you can assume you have introduced some technical error.** Let's calculate yours.

6. Click on a cell in the SD column. Type '=', this will open the formula dialog. Type 'STDEV' and select the correct formula from the drop down menu. Then click on the cell with the Ct value you want to calculate the SD of. Holding the ctrl button down (command on Mac), click its duplicate and press enter

7. Click on the SD cell you just calculated. Click on the little purple square and drag it down for the other values you want to calculate the SD for
   - ✓ In an ideal world, how far apart did you expect your duplicates to be?
   - ✓ How many of your duplicates have an SD>0.5? <u>Use your answers to fill out the results section</u>

8. Let's build a standard curve using your standards. First we have to turn the copy number of each your standards into the logarithm (base 10), so that we can compare it to the data obtained from the machine
   - ✓ What is, by definition, the $Log_{10}10^3$ (1 pt)?
   - ✓ $Log_{10}10^4$?

9. Fill in the the 'Log copy number' column for your standard samples

10. Now, we are going to build a chart to correlate the Log copy number with the Ct values read by the real time machine
    - ✓ What relationship would you expect between Log copy number and Ct values, if you performed and loaded your dilutions correctly (look at last week's theory)?
    - ✓ Do you recall what goes on the x axis?

11. Select the Log copy number data, then press ctrl (command on a mac) and select the column with the average Ct data you just calculated

12. Release the ctrl (or command) button and click
    - ✓ 'Charts'
    - ✓ 'Insert charts'
    - ✓ Scroll down to scatter and select 'scatter plot'. Your graph should look similar to the one in the figure below

 DO NOT CHOOSE LINE GRAPH! You are hoping your samples will sit on a line, but you don't know that yet. Choosing line graph would result in a biased regression curve and affect your statistical analysis

- ✓ What standard range was used to generate the graph in the figure below (2 pts)?

- ✓ Add pretty titles to allow easy interpretation of your data by tinkering with the format window up on the right (I'll show you or if they updated the program AGAIN we'll figure it out together, lol.)

- ✓ Is your standard curve linear?

- ✓ Do you recall what's the slope of a perfect real time standard curve (2 pts)?

- ✓ Would the slope be the same, if the dilution factor among standards were 4 (3 pts)?

13. We have to write an equation to describe your standard curve. Excel can do that for you. On the graph, select one of the data points by clicking it, then right click on it and select 'add trendline'
14. Select 'linear' then click on 'options'
15. In options select 'display equation on chart' and 'display R^2 value'. Click ok
16. Click on the equation to move it to the side of the chart. It should look like this (with your own values of course!)

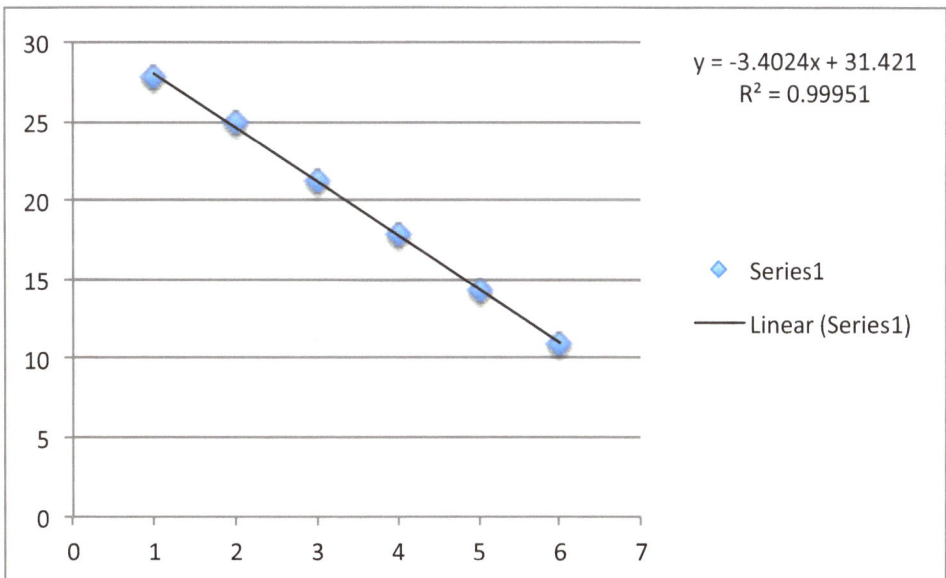

- ✓ What is the slope of the curve in the next page (1 pt)?

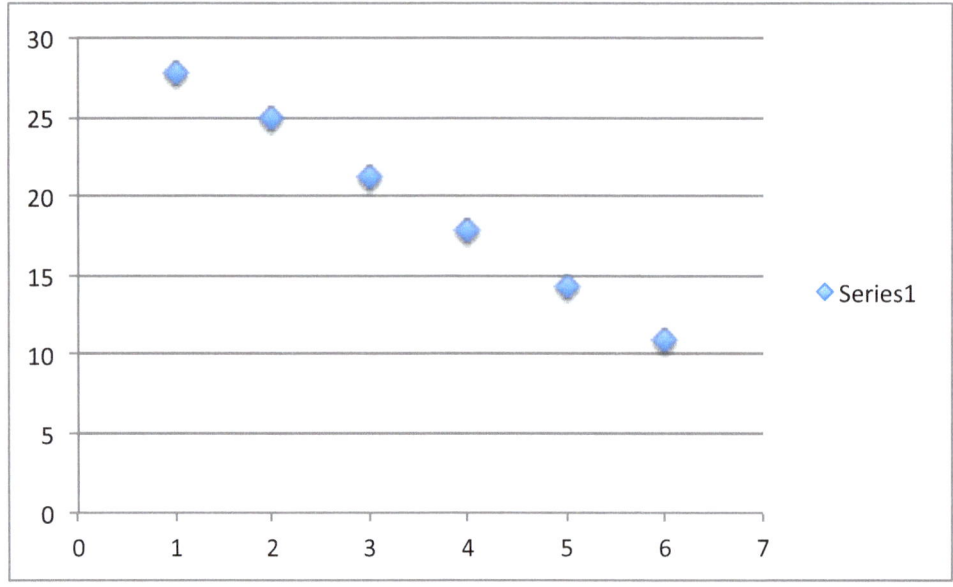

- ✓ What is the slope of your curve?

- ✓ The formula to calculate the efficiency of a primer set in real time is efficiency = $10^{(-1/slope)}-1$. What is the efficiency of the curve above (1 pt)?

- ✓ I know this primer set is 98% efficient. How efficient was it in your hands? <u>Write it also in your results</u>

- ✓ Is your efficiency acceptable (an acceptable range, for this lab, is 90-110%)?

17. The R^2 value gives you an idea of how well your data fit a linear regression. An R^2 value of 1 means a 100% fit, which is what we expect given we diluted the standards to fit on a perfect curve

    - ✓ What is the fit of the curve above (1 pt)?

    - ✓ What is the fit of your curve? Write it also in your results
    - ✓ Is it acceptable? (An acceptable fit has an r^2 value of 99-100%)

18. Now, we are going to calculate the levels of expression of GAPDH in your buccal cells
    - ✓ First of all, look at the Ct value of your cDNA. Does it fall within the range covered by your standard curve?

✓ Why is that important (2 pts)?

If your sample does not fall within the range of your standard use the Ct value of a classmate or make one up to go through the rest of the analysis. Just make sure to note this on your report. (Something like, "Since my Ct value did not fall within the range of my standards, I chose a fictional Ct value for educational purpose, just to see how to carry out the rest of the analysis.)

19. Look at the equation on your graph. You want to find the copy number of GAPDH in your cDNA

   ✓ Are you looking for x or for y? Do you have x or y as your data (3 pts)?

   ✓ Rewrite your equation in a format that is useful to calculate copy number of GAPDH

   ✓ Substitute your data into the equation. What number did you obtain?

   ✓ Is this the copy number or what (2 pts)?

   ✓ Calculate GAPDH's copy number in your real-time well

To be able to come up with an absolute value we have to relate the copy number we found to how much RNA was in the sample because, obviously, the more RNA you started from, the higher the cDNA, the higher the copy number of your transcript.

   ✓ Which experiments would you have done to get from RNA to real-time? (2 pts)

Knowing that:
- You would have added 10.8 μL of RNA to your retro-transcription
- You added 5 μL of sample to each real-time well (be it water, a standard, or cDNA)
- Assuming RNA was retro-transcribed with a 100% efficiency (1:1)
- Assuming your RNA concentration was 3.4×10^4 ng/mL

   ✓ Express your results as copies GPDH/ng initial RNA (5 pts)

Results, discussion, and conclusions

PROJECT 1

Presence of DNA precipitate (Y/N):

Presence of DNA pellet (Y/N)

DNA concentration (with unit):

260/280 ratio

(Discussion) What do you conclude from the ratio above?

260/230 ratio

(Discussion) What do you conclude from the ratio above?

DNA total yield (with unit):

(Discussion) If no DNA was obtained, hypothesize why:

h-TAS2R38-DNA-F:
h-TAS2R38-DNA-R:

DNA amount added to PCR sample (in ng):

Gel electrophoresis of TAS2R38 PCR (label the samples!):

The ladder:
Positive control (pos or neg):
Negative control (pos or neg):
Your sample (pos or neg):
Other samples on the gel (pos or neg)
Conclusions (remember to comment on the size of the PCR fragments):

DNA concentration after TAS2R38 PCR purification and quantification:

Your genotype at TAS2R38 as determined by sequencing (Tt, TT or tt):

Ability to taste PTC paper (Y/N):

Genotypic frequencies of class obtained by sequencing: TT =
    Tt =
    tt =

OBSERVED allelic frequencies of class obtained by sequencing: T = p =
    t = q =

Phenotypic frequencies of class obtained by sequencing: Tasters =
    Non tasters =

PREDICTED allelic frequencies obtained by HW: T = p =
    t = q =

Conclusions and discussion (check the objective of project 1 and write if you obtained it and what you found. Then write if most people in the lab obtained it and what were the general results. If you did not obtain some or all results discuss what you think went wrong:

PROJECT 2 (simulation)

RNA concentration: 3.4×10^4 ng/mL	
GAPDH primers: h-GAPDH-RNA-F: h-GAPDH-RNA-R:	
Size of intron included:	
Comments on specificity:	

Attach a picture of your real-time PCR curves (the amplification plot)

Attach a picture of your melting curves:

Attach your excel data (Ct values, means and SDs)

Samples with multiple Tm values:

Ct values of blanks:

Technical duplicates with a SD above 0.5 cycles:

Attach below a picture of your standard curve with the regression line and the r^2 value

r^2 value of your standard curve:

Efficiency of your standard curve:

GAPDH transcript copies/well:

Copies of the GAPDH transcript per ng of RNA:

Real-time data analysis and discussion

—Expected Results (ideal): in a perfect experiment, what did you expect for your "blank" values? What would you expect for your SD? How many cycles apart would the Ct value of your standard be? What would be the r_2 value? The efficiency? How many Tm values would you expect per each sample?

—Expected results (acceptable): alas, real results are never perfect! In an acceptable experiment, what would you expect for your "blank" values? What would you expect for your SD? How many cycles apart would the Ct value of your standard be? What would be the r_2 value? The efficiency? How many Tm values would you expect per each sample? The Ct value of the cDNA samples would have to be included in which Ct range to allow for analysis?

—Observed Results discussion: compare your expectations with the real data you analyzed. How did they deviate from the expectations? Based on your analysis, can you pinpoint where errors were made? Justify your answers by thorough explanations.

—Final analysis: the final result was copy number/ng RNA. Why? Explain this normalization as thoroughly and simply as you can. What mathematical steps did we have to go through to find that answer?

The cost of molecular science

It is so much fun to work in the lab that sometimes we lose perspective of the cost of the materials we use. This appendix is meant to put things into perspective and report the cost, as of January 2014, of just a few of the reagents we used this semester:

Filter tips, 10 boxes	$142.00
Taq polymerase, 250 units	$ 198.00
Trizol, 100 ml	$ 164.00
DNA ladder 100 bp, 100 applications	$ 112.00
RT-PCR kit, 50 reactions	$ 235.00
Real-time PCR SYBR2x, 500 reactions	$ 326.63
Real-time plates, 20	$ 58.00

Also, as you might have noticed, we did use a lot of plastic in the lab.
Whenever working in the lab, just like in your daily life, always keep in mind the environment; reduce the use of plastic whenever possible, reuse any plastic container that can be properly cleaned and recycle plastic whenever reusing is nor possible[9].
Also, all toxic reagents in the lab like ethidium bromide and phenol, are collected separately and disposed of according to the best environmental practices.

You are getting a degree in higher education.
If you don't set the good example, who will?

References

1. McLaren, A. Cloning: pathways to a pluripotent future. *Science* **288**, 1775-1780 (2000).
2. Gonzalez-Porta, M., Frankish, A., Rung, J., Harrow, J. & Brazma, A. Transcriptome analysis of human tissues and cell lines reveals one dominant transcript per gene. *Genome Biol* **14**, R70 (2013).
3. Prodi, D. A. et al. Bitter taste study in a sardinian genetic isolate supports the association of phenylthiocarbamide sensitivity to the TAS2R38 bitter receptor gene. *Chem Senses* **29**, 697-702 (2004).
4. Bartlett, J. M. & Stirling, D. A short history of the polymerase chain reaction. *Methods Mol Biol* **226**, 3-6 (2003).
5. Saiki, R. K. et al. Primer-directed enzymatic amplification of DNA with a thermostable DNA polymerase. *Science* **239**, 487-491 (1988).
6. Sanger, F., Nicklen, S. & Coulson, A. R. DNA sequencing with chain-terminating inhibitors. *Proc Natl Acad Sci U S A* **74**, 5463-5467 (1977).
7. Chomczynski, P. & Sacchi, N. The single-step method of RNA isolation by acid guanidinium thiocyanate-phenol-chloroform extraction: twenty-something years on. *Nat Protoc* **1**, 581-585 (2006).
8. Higuchi, R., Fockler, C., Dollinger, G. & Watson, R. Kinetic PCR analysis: real-time monitoring of DNA amplification reactions. *Biotechnology (N Y)* **11**, 1026-1030 (1993).
9. Bistulfi, G. Sustainability: Reduce, reuse and recycle lab waste. *Nature* **502(7470)**, 170 (2013).